CUSTODIANS OF THE COMMONS

The United Nations Research Institute for Social Development (UNRISD) is an autonomous agency engaging in multi-disciplinary research on the social dimensions of contemporary development problems. Its work is guided by the conviction that, for effective development policies to be formulated, an understanding of the social and political context is crucial. The Institute attempts to provide governments, development agencies, grassroots organisations and scholars with a better understanding of how development policies and processes of economic, social and environmental change affect different social groups. Working through an extensive network of national research centres, UNRISD aims to promote original research and strengthen research capacity in developing countries.

Current research themes include: The Challenge of Rebuilding War-Torn Societies; Integrating Gender into Development Policy; Environment, Sustainable Development and Social Change; Crisis, Adjustment and Social Change; and Volunteer Action and Local Democracy: A Partnership for a Better Urban Future. New research is beginning on: Follow-Up to the Social Summit; Business Responsibility for Sustainable Development; New Information and Communications Technologies; Culture and Development; Gender, Poverty and Well-Being; and Public Sector Reform and Crisis-Ridden States. Recent research programmes have included: Ethnic Conflict and Development; Socio-Economic and Political Consequences of the International Trade in Illicit Drugs; Political Violence and Social Movements; and Participation and Changes in Property Relations in Communist and Post-Communist Societies. UNRISD research projects focused on the 1995 World Summit for Social Development included Rethinking Social Development in the 1990s; Economic Restructuring and Social Policy; Ethnic Diversity and Public Policies; and Social Integration at the Grassroots: The Urban Dimension.

A list of the Institute's free and priced publications can be obtained by writing to: UNRISD, Reference Centre, Palais des Nations, CH-1211, Geneva 10, Switzerland. Fax: (41 22) 740 0791; e-mail: info@unrisd.org; World Wide Web Site: http://www.unrisd.org.

CUSTODIANS OF THE COMMONS
Pastoral Land Tenure in East and West Africa

edited by

Charles R Lane

Earthscan Publications Ltd, London

First published in the UK in 1998 by
Earthscan Publications Ltd

Copyright © International Institute for Environment and Development (IIED), 1998

All rights reserved

A catalogue record for this book is available from the British Library

ISBN: 1 85383 473 4

Typesetting and page design by Derbyshire Design

Printed and bound by Biddles Ltd., Guildford and Kings Lynn

Cover design by Yvonne Booth

For a full list of publications, please contact
Earthscan Publications Ltd
120 Pentonville Road
London N1 9JN
Tel: 0171 278 0433
Fax: 0171 278 1142
email: earthinfo@earthscan.co.uk
http://www.earthscan.co.uk

Earthscan is an editorially independent subsidiary of Kogan Page Ltd and publishes in association with WWF-UK and the International Institute for Environment and Development.

TABLE OF CONTENTS

About the Authors   *vi*
Preface   *ix*
Acknowledgements   *xii*
List of Maps   *xiv*
List of Tables   *xv*
List of Abbreviations   *xvi*

1	Introduction: Overview of the Pastoral Problematic	1
2	Kenya	26
3	Mali	46
4	Mauritania	71
5	Senegal	93
6	Sudan	120
7	Tanzania	150
8	Uganda	169

Appendices   *187*
References   *208*
Index   *225*

ABOUT THE AUTHORS

Mohammed Abusin received his PhD from the University of London. He is currently Professor of Geography at the University of Khartoum, and has published widely on population dynamics, pastoralism, development and environment.

Aboubakrim Deme was born into an important maraboutic family in northern Senegal, and trained as an agricultural engineer. He has worked throughout Senegal for a variety of parastatals as well as in neighbouring countries such as Mali and Tchad on international projects. On his retirement in 1982, he was elected President of the Association for the Renaissance of Pulaar, and began a second career promoting the development of non-formal education in national languages. In 1989, he was a founding member of the Groupe d'Initiative pour la Promotion du Livre en Langues Nationales, and has worked tirelessly to translate the results of pertinent research on herding, farming and environmental issues into his own first language, Pulaar. He is the author of several original works in Pulaar, including collections of oral traditions, and a thematic Pulaar–Pulaar dictionary on herding, farming and environmental vocabulary.

Amadou Tamsir Diop is an agro-pastoralist veterinarian. In 1983, after three years' experience in the Livestock Department at national and regional level, he joined the Senegalese Agricultural Research Institute as researcher before becoming head of programme at the Dakar-Hann National Livestock and Veterinary Research Laboratory. He has been working for several years with various development projects and the national agricultural extension programme. Since 1994, he has been employed at the Dahra-Djoloff Animal Science Research Centre, where he was appointed leading researcher in October 1995.

Alioune Ka is a pastoralist specialising in remote sensing techniques. Between 1979 and 1989, he was in charge of the Pastoral Service of the East Senegal Livestock Development Project. He is currently working as pastoral resources expert at the Dakar Environmental Monitoring Centre in Senegal, where he also trains technicians in pastoral monitoring.

About the Authors

Wilberforce Kisamba-Mugerwa is both Minister of State in the Ugandan government, and a Senior Research Fellow at the Makerere Institute for Social Research at Makerere University. He has written widely on land tenure issues and has an informed interest in pastoral studies.

Mukhisa Kituyi is a pastoral scholar whose doctoral thesis on the Kenya Maasai was published under the title *Becoming Kenyans*. He later developed a pastoral studies programme at the Africa Centre for Technology Studies in Kenya. He has more recently become an opposition Minister of Parliament and sponsored legislation on pastoralist issues. At the same time he has undertaken a number of consultancies in pastoralist related fields.

Charles Lane is a social scientist who has worked in East Africa for ten years and wrote his doctoral thesis on Barabaig resource tenure in Tanzania which has been updated and published under the title *Pastures Lost*. Until recently he was Senior Research Associate at the International Institute for Environment and Development where he worked with pastoral organisations on a pastoral land tenure research and policy programme in Africa. He is currently acting as an independent consultant and working with Survival International on a programme of support for pastoralists to successfully assert their rights to land in East Africa, and is a director of Pilotlight, a new organisation which works to address the issue of landlessness worldwide.

Richard Moorehead is a political scientist who worked in the inland Niger delta in Mali for seven years and wrote his doctoral thesis on customary and contemporary resource tenure systems in the area. He is presently a Research Associate at the International Institute for Environment and Development, working on resource tenure issues, gender, poverty, national environmental policy and community-based natural resource management systems, mostly in francophone West Africa.

Daniel Ndagala studied sociology and economics at the University of East Africa, and later Dar es Salaam. His fieldwork has focused on nomadic and pastoral peoples in Tanzania. He has published books on the Ilparakuyu and Maasai, and numerous papers on rural development policy for pastoralist and hunter-and-gatherer societies. As Commissioner for Culture he has most recently been concerned with pastoral land tenure issues.

Ibrahima Niang is a veterinarian who graduated from the University of Dakar in 1982. In 1983 he became the Head of the Research and Economic Planning

Department at the Ministry of Livestock, coordinating the Permanent Diagnostics Programme between 1986 and 1990. Since 1990, he has been Head of the Pastoral Department, responsible for pastoralists' training programmes.

Mohamed Ould Zeidane studied applied mathematics in the Ivory Coast and then in France. From 1988 to 1991 he was Head of Pastoral Studies within the Livestock Department of the Ministry of Rural Development in Mauritania. Since then he has been Economic and Rural Affairs Advisor to the Prime Minister. He is also a specialist and international consultant in livestock, and commercialisation of livestock production, agricultural economics, land tenure and rangeland management policies. He is currently involved with monitoring pastoral associations and rangeland natural resource management for the World Bank.

PREFACE

Is pastoralism a dying tradition? Is it obsolete, inefficient and environmentally damaging? Should the pastoralists of Africa be persuaded to settle down and engage in activities more relevant to the economic realities of the 21st century?

This important volume gives a resounding 'no' to these questions. With contributions from some of the most knowledgeable and dedicated people studying pastoralist issues today, it documents the importance of pastoralism in the countries in which it is practised. Pastoralism provides both a living and a way of life to over 25 million people in Africa, and makes a significant contribution to the economies of the region. It is a production system that continues to make more efficient and sustainable use of the dryland resources of Africa than most alternatives that have been tried.

Yet the investment in this sector has seldom been proportionate to its economic contribution; pastoralists are far more likely to be ignored or discriminated against than to be provided with support. This lack of support has had adverse impacts on economic growth, social development, environmental sustainability and human rights. Pastoralists are suffering disproportionately in Africa, with many groups experiencing declining living standards and increased poverty and insecurity. In part, this is because pastoralism is an inherently risky way of life, relying on scarce resources and suffering from the effects of recurring drought. But in many areas these difficulties have been compounded by the fact that the best traditional common grazing lands have been privatized or allocated to agricultural schemes, and the land that has been retained in pastoralist production has therefore undergone degradation.

The most common response to these problems is to attempt to settle pastoralists. Land titling, land use planning, privatization, fencing, and settlement schemes have all been promoted in pastoralist areas. The authors in this volume, however, argue that such schemes have not furthered the cause of social development. Settled pastoralists are more likely than traditional pastoralist groups, and more likely than the general population, to suffer from extreme poverty and social dysfunction. Settlement schemes typically increase social inequality, and increase the burdens of women whilst depriving them of customary rights.

Pastoralism has been misunderstood since colonial times. Although it can appear to be a relatively unproductive use of resources, especially of more fertile lands, in fact pastoralism often makes the best use of the harsh environmental conditions of Africa's drylands. Typically, traditional pastorialist production systems cover large areas of relatively unproductive drylands as well as smaller areas that are wetter and more fertile. It is the use of these wetter areas to sustain the herds through the dry season that allows pastoralists to make use of the drylands in the rest of the year. Efficient use of the drylands depends on pastoralists' ability to move herds away from them during the driest periods of the year before they become degraded. The nature of pastoralist production systems thus depends on movement, and a relatively non-intensive use of the best land is necessary in order to make any use at all of poorer lands. If, as is commonly the case, the most productive lands are the first to be converted to farmlands, and if the mobility of pastoralists is curtailed, the production per unit of wetland may also rise. However, production in the surrounding drylands will almost certainly fall, and environmental degradation will be the likely result.

This volume documents the effectiveness of traditional pastoral production systems. It is, however, a book that looks forward rather than back. The authors recognize that it is impossible to return to the traditional sustainable management practices of the past. Population increases, alienation of land, restrictions on migratory movements and a decline in rainfall in pastoralist areas have all made pastoralism more difficult to sustain. In addition, the social relations that established and maintained traditional pastoralist practices have changed. Social differentiation within pastoralist groups has increased—in fact, today many livestock holders are not traditional pastoralists at all, but are absentee herders who acquired animals during times of drought and maintain their residences in urban areas. In addition, with population movements and increasing ethnic diversity in pastoralist areas, the question of which 'tradition' to follow is itself unclear.

The success of pastoralism depends on efficient systems of resource management, but the traditional practices that maintained such systems in the past cannot be reproduced. How can this dilemma be resolved? Recognizing that pastoralism is a viable and productive use of resources must be the first step. Establishing a better understanding of ecological conditions in areas in which it is practised is also necessary, as is a careful examination of changing tenure arrangements and their effects, not just on pastures converted to farmlands, but on the whole pastoralist system. In addition, attention must be paid to establishing mechanisms for conflict resolution in these areas of increasingly complex social institutions. This book argues that organization of pastoralists is a key to improving the viability of pastoralist production systems. While traditional practices must change, rangeland management is too complex an undertaking to be codified by outsiders, and pastoralists must be involved in determining the shape of rangeland management in the future.

The case studies that make up this report were commissioned by the International Institute for Environment and Development (IIED), and were supported by UNRISD's research programme on Environment, Sustainable Development and Social Change.

Jessica Vivian
on behalf of UNRISD
June 1997

ACKNOWLEDGEMENTS

Following the workshop: 'Sustainable Development Through People's Participation in Resource Management' at the United Nations Research Institute for Social Development (UNRISD) in Geneva in May 1990, it was agreed by UNRISD to support the International Institute for Environment and Development (IIED) to commission a series of pastoral land tenure overviews from selected African countries. Guidelines were sent to authors who produced most of the text for this book (see Appendix I).

A project involving detailed correspondence with distant foreign countries can only succeed with the help and forbearance of many people. I am particularly grateful to the authors who responded favourably to my request and provided such rich and diverse material. It is hoped that seeing their work in print will go some way to compensating them for the long delay in its publication.

Working with issues on both sides of the continent caused me to rely on colleagues more familiar than myself with some of the countries involved. The Pastoral Network for the Horn of Africa (PENHA) provided names for contributors in the region. My colleague at IIED, Camilla Toulmin, helped me identify some of the authors in West Africa, and she and Ced Hesse commented on their submissions. Camilla also provided comments on the Introduction to this text. Other colleagues at IIED, Nicole Kenton, Judy Longbottom and Aleth Abadie, assisted me with most efficient secretarial and administrative support throughout the life of the project.

A number of people provided important specialist skills for the preparation of this book. Jean Lubbock translated texts from French into English. Kemal Mustafa and Alan Scholefield made substantial contributions by editing some of the papers in English, and Kemal helped me identify some common themes for consideration when dealing with the text as a whole. Caroline White provided expert professional copy editing of the final text. Map 1 was ably prepared by Tim Aspden of the Geography Department at University College, London.

Much of the material presented here would not have come to light without the support of UNRISD. I am particularly indebted to Dharam Ghai and Jessica Vivian who recognised the importance of clarifying the issues related to pastoral land tenure for attaining sustainable development with pastoralists throughout the continent. From within UNRISD I also benefitted from the support of Adrienne Cruz in the production of the manuscript, and Wendy Salvo and Rosemary Max for administration of the project.

Charles R Lane
IIED
April 1997

LIST OF MAPS

Map 1	Countries in Africa Referred to in Text	xviii
Map 2	Kenya	26
Map 3	The Two Maasai Districts of Kenya	27
Map 4	The Turkana District of Kenya	28
Map 5	Dispersal of Population in Northwest Turkana	42
Map 6	The Republic of Mali	46
Map 7	The Inland Delta of the River Niger	47
Map 8	The Seasonal Flood Regime of the Inland Niger Delta	48
Map 9	The Islamic Republic of Mauritania	71
Map 10	Senegal	93
Map 11	Ecological and Administrative Regions of Senegal	97
Map 12	Sudan	120
Map 13	Sudan Ecological Zones	125
Map 14	Climatic Regions of Sudan	126
Map 15	Soil Regions of Sudan	127
Map 16	Vegetation and Rainfall of Sudan	127
Map 17	Land Use in Sudan	128
Map 18	Major Areas of Communal Grazing in Sudan	129
Map 19	The Camel Environment in Sudan	130
Map 20	Environmental Degradation in Pastoral Homeland Sudan	142
Map 21	Nomadism in Blue Nile Province, 1955	144
Map 22	Agricultural Schemes in Blue Nile Province, 1986	145
Map 23	Tanzania	150
Map 24	Area of Major Land Alienation in Tanzania's Pastoral Areas	152
Map 25	Uganda	169
Map 26	The Cattle Corridor in Uganda	171

LIST OF TABLES

Table 5.1	Funding of Development Projects in Senegal, 1988–1991	99
Table 5.2	Agro-sylvo-pastoral Developments in Senegal between 1960 and 1990	109
Table 6.1	Nomadic Population by Regions, 1983 Census	121
Table 6.2	Livestock Numbers by Province, 1981/82	121
Table 6.3	Livestock Population Estimates (in millions), 1917–2000	123

LIST OF ABBREVIATIONS

ANEM	National Association of Mauritanian Herders
CIDA	Canadian International Development Agency
DDC	District Development Corporation
EU	European Union
FAO	Food and Agricultural Organisation
GEI	Groupements d'Interet Economique (Economic Interest Groups)
GIO	German Imperial Ordinance
IDA	International Development Assistance
IIED	International Institute for Environment and Development
IUCN	World Conservation Union
KIPOC	Korongoro Integrated People Oriented Conservation
MISR	Makerere Institute of Social Research
NARCO	National Ranching Company
NCA	Ngorogoro Conservation Area
NGO	Non-governmental organisation
NPAs	Nomadic Pastoral Associations
NOPA	Nomadic Pastoralists in Africa
NORAD	Norwegian Agency for Development Cooperation
ODEM	Operation pour le Developpement de l'Elevage dans la Region de Mopti (livestock parastatal)
ODR	Operations de Developpement Rurale (parastatal development agencies)
PAs	Pastoral Associations
PDESO	Livestock Development Project in Eastern Senegal
PENHA	Pastoral Network for the Horn of Africa
PSAESAS	Senegalo-German Agro-sylvopastoral Project
PTIP	Public Investment Programme
SAED	Exploitation Agency for the River Delta
SODESP	Livestock Development Agency for the Sylvopastoral Area
SRDRs	Regional Rural Development Agencies
UBT	Unite de Betail Tropical (Tropical Livestock Unit)
UNICEF	United Nations Children's Fund
UNRISD	United Nations Research Institute for Social Development
USAID	United States Agency for International Development
UN	United Nations
VPAs	Village Pastoral Associations

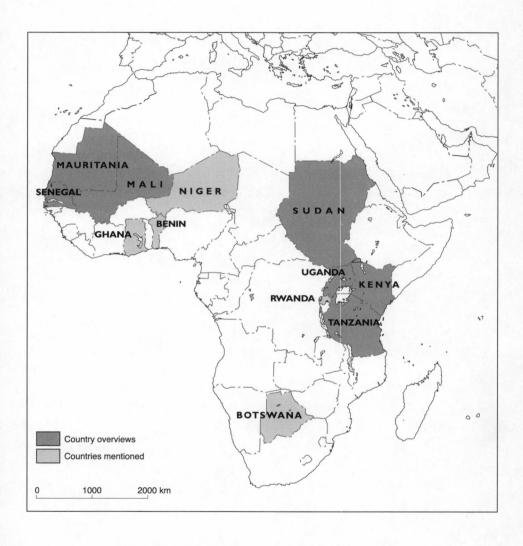

Map 1 Countries in Africa Referred to in Text

1

INTRODUCTION

Overview of the Pastoral Problematic

...land belongs to a vast family of which many are dead, few are living and countless members are still unborn. (Nigerian herder)

This herder's statement captures the essential characteristic of pastoral land tenure in Africa. Land belongs to a group or 'family' that is linked by descent or cultural affiliation. It is not 'owned' in the sense that users enjoy unlimited rights to exploit and dispose of it at will. It is held in trust by the living for future generations. To ensure that they inherit land currently enjoyed by the living, levels of use are limited by the right of usufruct—the right to enjoy the product of land only in so far as it does not cause damage and reduce its future productive capacity. It is in this way that pastoralists are custodians of the commons.

Today this concept of land is little understood and even less respected by many African governments and Western donors. The result has been the overriding of customary pastoral land tenure systems creating land tenure insecurity to the general disadvantage of pastoral peoples. Lands once managed sustainably have been alienated and have often become degraded by inappropriate forms of use.

Not surprisingly throughout Africa land is the issue that concerns pastoralists most. It seems that a whole land user group is unified in its anxiety over a single problem it is experiencing. What is the problem of land for pastoralists, and why is it they seem to be suffering so disproportionately?

Before we can provide some answers to these questions we need to understand more about how pastoralists organise themselves in relation to land and find out what distinguishes them from other land users. We also need to know what processes are going on in pastoral areas of Africa that affect pastoralists and their enjoyment of land. In providing these answers it is hoped to reveal how pastoralists can become more effective in protecting and managing their lands within the new constraints they face.

The International Institute for Environment and Development and the United Nations Research Institute for Social Development together decided to explore the issue of pastoral land tenure and its impact on resource management and environmental degradation in pastoral areas of Africa. Their interest was prompted by research findings that suggest changes taking place on Africa's rangelands are undermining traditional pastoral livelihoods. While there may be many causes for this, the purpose of commissioning a series of overview studies has been to explore the relationship between land tenure and these changes.

Case studies were selected from seven different African countries—both anglophone and francophone—with sizable pastoral populations: Kenya, Mali, Mauritania, Senegal, Sudan, Tanzania and Uganda (see Map 1). The overall aim of the project is to inform decision-makers about the problem of land tenure for African pastoralists, and identify ways to stem the decline and attain greater land tenure security for pastoral peoples in the future. Each author was asked to compile an overview of the problem of pastoral land tenure in a selected country so as to:

1. design alternative policy interventions;
2. support pastoralists and their organisations better to deal with the problem of land;
3. guide future research in the field of pastoral land tenure (see Appendix I for terms of reference).

This introductory chapter sets out the case material in a broader context by not only making use of material offered by contributing authors but also by drawing on examples from different countries and other sources. Of particular importance is the review of material from East (anglophone) and West (francophone) Africa where the colonial legacies differ, but problems for pastoralists seem to have much in common. It is hoped that a comparison of the issues in the two regions will enable us to highlight important differences and build on common themes. Through this it is hoped to make connections between the two regions and help pastoralists and those working with them to share perspectives and benefit from each other's work.

A RAPIDLY CHANGING CONTEXT

Africa's pastoralists tend to occupy dryland environments and have always suffered recurring droughts and times of food shortage. However, despite a relative scarcity of natural resources and the vagaries of climate they were able to survive. This was achieved in part by migratory land use and systems of communal land tenure which together enabled them to cope during periods of climatic stress. Their variable use of resources created and maintained much of Africa's savanna, and until recently this practice had little adverse impact on the environment. Today, pastoralists are finding it more difficult to cope, partly because of population increases relative to the availability of resources, partly because of climatic conditions that show an overall decline in rainfall in the West and Central Sahelian zone over the last 30 years (Hulme 1992), but also because of restrictions on migratory movements and alienation of their lands. As a consequence they make up some of the most geographically and politically marginalised peoples in Africa.

Conditions for pastoralists in recent years have worsened considerably. Ever increasing areas that were once communal pastures have been lost to pastoral production. Irrigation schemes, smallscale farming and mechanised agriculture have withdrawn large tracts of the most productive land for non-pastoral use. Food production per head and living standards for pastoralists have fallen. Future incomes and welfare are further threatened by increased degradation of land while a growing conflict of interests is pitting pastoral communities against governments, against each other, and against other land users.

These conflicts do not usually manifest themselves as largescale armed conflicts although this has sometimes been the case, particularly in West Africa; they are more structural and hidden (Bradbury, Fisher and Lane 1995). They are nonetheless violent in their impact of forcing people from their homes and in the violations of human rights which can follow from evictions, as has recently been experienced in Kenya.

Historically, pastoral groups have managed conflicts over resources through tried and tested traditional systems. However, with tenure reform and the alienation of pastoralists from their lands, customary methods of negotiation, arbitration and adjudication are breaking down in competition with more omnipotent forces.

The rapid transformation of pastoralism is increasingly shifting control over land to small, male-dominated elites, some from within pastoral societies but mostly from agricultural, urban, and civil service or military backgrounds. The vast majority of herders, impoverished and politically deprived, find themselves with neither enough animals nor sufficient access to rangeland and water to sustain their livelihoods. Under these conditions men become either hired herders for absentee herd owners, migrate out of pastoral areas for waged labour, while women remain at home, assuming greater responsibility for the management of herds and flocks, or are expected to fend for themselves without any animals for as long as men are away.

It is becoming more widely understood that pastoral women have an important role in resource management. Many pastoral societies, for example, have specific social institutions through which women exert their authority over access to resources. However, it is also becoming clear that women suffer adverse impacts in their role as resource managers when tenure is reformed as their rights of access and control over land and water resources is undermined. There is also some evidence to suggest that women's vulnerability is compounded by the progressive alienation of rangelands away from community-based ownership structures, and as privatisation tends to break up labour groups women, particularly in female-headed households, lose access to labour pools. Thus women end up suffering from a triple jeopardy: loss of animals, loss of labour and loss of land (Horowitz and Jowkar 1992).

All seven case studies presented here provide evidence of the acceptance of outdated policies that guide pastoral development. Pastoralists are still thought incapable of efficient land use and independent governments have willingly adopted colonial measures that favour agricultural production and the alienation of pastoral lands for non-pastoral use. The focus has been on the use of legislation to reform land tenure rather than accommodate the different pastoral land tenure systems found in each locality. National legal frameworks provide the essential conditions within which tenure systems operate, but they are inconsistent with customary land tenure arrangements that enable pastoralists to make good use of natural resources.

The studies do, however, suggest other factors that influence the way pastoral land tenure systems operate in the wider context, for example: by referring to the significant contribution pastoralists make to national economies; the prejudices against pastoralists that exist within national political systems; their relative powerlessness in relation to the nation state; the technical nature of development interventions; and problems of land degradation which are taken up in each case study according to their relevance to each country.

The issue of pastoral land tenure is important for no better reason than its scale. Pastoralists make up a significant proportion of Africa's population, numbering about 25 million people (Sandford 1983), constituting around 12 per cent of the total population of nearly 200 million people in countries of the dry belt of East and West Africa (Bonfiglioli 1992), and utilising around 500 million hectares of grassland (World Bank 1989).

The country overviews provide data confirming that, despite pastoral dependence on marginal arid and semi-arid lands subject to most difficult climatic conditions, they are major suppliers of meat, milk and hides for national and international markets. Empirical information is offered on the actual contribution pastoralists make to their national economies. Even relying on official statistics that do not take account of sizeable 'unofficial' economic activities it is clearly more significant than is generally realised or publicised. We learn, for example, that Tanzanian pastoralists and agro-pastoralists own 90 per cent of the 13 million cattle in the country. In Sudan, it is estimated that 20 per cent of the population are pastoralists who own over 80 per cent of the 45 million camels, cattle, sheep and goats, and that the livestock sector earns over 20 per cent of Sudan's foreign exchange. In Senegal, 12 per cent of the population are pastoralists who hold more than 50 per cent of the cattle, sheep and goats. In Mauritania, more than 50 per cent of the population depend on at least some of their income coming from about 10 million head of livestock, and pastoral production accounts for at least 20 per cent of gross domestic product (GDP). The economic significance of pastoral production in the countries studied is therefore of a magnitude which has rarely been recognised by national administrations. The case studies highlight the difference between this contribution and efforts to promote pastoral production.

THE TOP–DOWN APPROACH TO REFORM

Pastoralists form an important and distinct population, whose methods of production are based on livestock rearing and associated transhumance. However, as the case studies show, governments and donors have consistently failed to take this into account when designing development interventions in rangeland areas. Development interventions are too often based on the premise that pastoralists accumulate numbers of livestock beyond economic requirements, and that their land tenure systems are structurally incapable of efficient land use and will inevitably lead to land degradation through overgrazing. It is argued that the way pastoralists hold land in 'common' prevents them from attaining high levels of commercial offtake, limiting their stock numbers within the carrying capacity of land, and as a consequence failing to protect pastures from over-use. This is thought to be reason enough for reform of indigenous pastoral land tenure arrangements and the imposition of new administrative requirements and land laws.

The major policy emphasis has been on transforming pastoral production through technical interventions and the imposition of land tenure reform through measures for settlement, land titling and formal land use planning, privatisation of the commons, and the alienation of pastures for non-pastoral purposes. As is well illustrated in the case material offered in the following chapters, development efforts in the pastoral sector of Africa have almost universally failed to bring expected advances in livestock productivity, improvements in pastoralist welfare, or protected rangelands from degradation (Sandford 1983).

The 'old orthodoxy' (Lane and Swift 1989), or what Sandford (1983) calls the 'mainstream view' in which pastoralists are thought to keep livestock in numbers way beyond economic requirements and make use of land tenure systems that are thought to be structurally incapable of efficient land use, has been challenged and is now recognised as a flawed basis on which to design rangeland development strategies.

Despite research findings which consistently demonstrate the relative superiority in productivity per unit area of land compared with commercial ranches in comparable ecological conditions (Behnke 1985), traditional pastoral production and the customary land tenure arrangements that make it possible are still regarded as obstructions to development. However, these views and the policies they spawn still advocate the alienation of vast areas of pasture land for non-pastoral use by individuals and the state, for commercial agricultural production, or the conservation of wildlife.

Misconceptions about pastoralists and their production systems within independent states abound. Acceptance of the 'tragedy of the commons' argument has led to tacit government and donor support for the privatisation of pastoral commons. This in turn has facilitated encroachment by smallscale cultivating agriculturalists onto pastoral lands and the alienation of pastures for largescale commercial farming operations and ranches which has led to conflict over natural resources between pastoralists and cultivators, and between pluralists and the state (Bradbury, Fisher and Lane 1995). Where pastoralists have tried to maintain communal management of resources this has been undermined by state-sponsored and spontaneous privatisation within pastoral society of available resources.

CHANGES IN LAND TENURE

For a thorough understanding of pastoral land tenure issues it is important to review the theoretical considerations and central concepts that have been used in discussions surrounding pastoral land tenure in Africa.

Land tenure is defined as the 'terms and conditions on which natural resources are held and used' (Bruce 1986). It can be described as the manner in which pastoral resources are *owned*: ie the *property* relations that are sanctioned by the society in which people live. Property in this context is defined as:

> *...a claim to a benefit stream that some higher body—usually the state—will agree to protect. Property is not an object but is rather a social relation that defines the property holder with respect to something of value against all others. Property is a triadic social relation involving benefit streams, rights holders and duty bearers.* (Bromley et al 1992)

INTRODUCTION

Before any discussion of pastoral land tenure it is important to distinguish between property and that which is not property (Lane and Moorehead 1994). This is because of the widespread and continuing confusion that exists in the literature and in policy over 'open access' resources, which are *not* by definition owned by anyone and are not subject to tenure rules (and therefore are not property at all), and 'controlled access' resources which may be owned by several overlapping categories of rights holders.

All pastoral lands in Africa are held under controlled access tenure regimes, often communal in form. Communal tenure relates to that system of rights in which 'access to land [is] based upon membership in a group such as lineage...defined by common descent' (Bruce 1986). It is more complex than common property as it can include different classes of tenure ranging from true common property through to individual private property within the communal system.

Birgegard (1994) expands the definition of tenure beyond man's relationship to land, and argues that tenure is a social institution in which there is a relationship between individuals and groups which govern a series of rights and duties with respect to the use of land. As such it touches all aspects of life through its role in people's survival, the distribution of wealth, political power and cultural expression. This means that enforced changes in tenure are likely not only to alter the way people relate to land as an economic resource, but have a profound effect on the entire social fabric of society. As will be illustrated by the following case studies such effects are unpredictable and can have a destabilising influence on national as well as local affairs.

Broadly speaking, pastoral range resources in Africa are owned under three controlled access property regimes: as state (national property), communal property, and private property. Many pastoralists may use the whole range of these property types in pursuing their livelihoods (Lane and Moorehead 1994). As Behnke has pointed out, this type of tenure system can be conceived as incorporating all these property rights:

These tenure systems can be envisaged as a matrix in which rights to different resource categories are partitioned within a hierarchy of different ownership groups ranging from the individual producer up to the largest tribal or ethnic group. Mobility is possible because these ownership groups are not territorially distinct but possess overlapping and potentially conflicting rights to different categories of resources in one area. (1992)

Pastoral land tenure systems differ from Western models with more uniform individual title that can be sold in a market, but it does not mean that African communal land users have any less strong sense of ownership, nor customarily lack security of tenure with respect to their land (Bruce 1986).

In general terms, there are three prevailing economic models of African rangeland use and tenure that have influenced thinking on tenure in the pastoral context:

1. that which advocates privatisation of property under the assumption of Hardin's 'tragedy of the commons' theory (1968, 1988);
2. the 'property rights' school (Behnke 1985,1991);
3. a school of thought which has been termed the 'assurance problem' led by Runge (1981, 1984, 1986) and others which advocates common property resource management (Bromley and Cernea 1989).

All three models are based on simple and persuasive theories about the relationship between natural resources and how they are used by rural land users. None of them is free from ideology, yet they are presented as truths despite inadequate empirical testing and rigorous evaluation (Lane and Moorehead 1994).

The tragedy of the commons theory argues that indigenous communal land tenure systems are incapable of efficient land use as herders with the means and desire to expand their livestock holdings will ultimately destroy the range through overgrazing. As each individual has unrestricted access to the common and is motivated by a wish to maximise his herd, so tragedy is inevitable. The logic that lies behind this is that for each animal a herder adds to his herd on a common, the extra consumption of resources will provide a direct benefit to him alone but the cost of that increased consumption will be shared by all. As the benefit enjoyed by the individual exceeds the shared cost, the herder is encouraged to continue increasing his herd size even if it results in the destruction of the range and a general loss for all. As such new rules have to be imposed on herders from outside society to control land use. This argument has been tremendously influential—amounting to dogma—and has been used to justify policies for tenure reform for the privatisation of land, the registration of title deeds and formal land use planning.

The property rights school argues that as resources become more valuable (due to market demand for the product of that resource, and increasing population pressure) they will become increasingly controlled. In these conditions herders can develop their own management institutions in line with the scarcity of resources. Hence, outside intervention is not required.

The assurance problem helps explain how in the past pastoral societies had effective management institutions that allocated access to resources between co-owners and excluded outsiders (through mutual assurance that members of society will respect rules and duties), and that these institutions have been undermined by wider political, social and economic factors emanating from outside herder societies. As a corollary, it is argued that sustainable management systems based upon appropriate tenure arrangements can be generated by herders themselves — but only if they are provided with adequate support and encouragement to do so.

There are four basic tenets of land tenure reform presently underway in Africa that are profoundly affecting pastoral lands and creating heightened land tenure insecurity for pastoral populations: the nationalisation of natural resources, sedentarisation of nomadic herders, titling and formal land use planning, and the privatisation of land.

Nationalisation

In the interests of nationhood and the development of the post-colonial state, governments with both anglophone and francophone traditions have accepted the colonial legacy of nationalised property in which radical title to land is held in the public domain and authority over its dispensation is entrusted to the head of state. The perceived inability of pastoralists to control the size of their herds and protect resources from destruction has legitimised the state's interest to assume ownership and management of land.

However, current research reveals that nationalisation of pastoral lands is creating very different outcomes to those expected (Lane and Moorehead 1994). It is becoming increasingly clear that nationalisation is breaking down customary land tenure arrangements and failing to replace them with an effective alternative management system.

The same is true of development interventions such as the provision of water on rangelands and the irrigation of river margins. Whereas before customary rights' holders controlled access to water, today 'public' water points are accessible to all. This has concentrated people and livestock in new locations and transformed traditional land use patterns that have led to environmental stress (Thebaud 1990). In Mauritania religion has played an important part in dissolving customary management systems and led to conflict as Koranic law provides much broader access to water and grazing than the customary systems (Zeidane 1993).

Ironically, nationalisation of land may well be creating the conditions for the 'tragedy of the commons' to take place. Where the state is unable to provide adequate management and yet at the same time insists that 'everyone' has a right of access by virtue of citizenship, the stage is set for individual herders to invest in more animals while ignoring the public cost of their actions, for the simple reason that if they do not make use of the pasture someone else will. Crucially, they no longer have any say in who the 'someone else' is, and can take no action to prevent their entry onto the range. Research is further showing that this is not as much a result of the state failing to manage, but interests prevailing in the maintenance of tenure ambiguities which allow elites to acquire access to pasture lands on a scale that would otherwise be impossible (Lane and Moorehead 1994).

Sedentarisation

It is a simple step in logic to extend the 'tragedy of the commons' argument from one of believing in the inability of pastoralists to manage land effectively, to believing that mobile land use is evidence of disorganised lives, and from there to impose policies for the settlement of pastoralists. This strategy constitutes perhaps the greatest single transformation of pastoralism as a means of production and way of life. Despite the inherent contradiction of settling people who rely on mobility to exploit range resources, and the subsequent adverse impact on pastoralists and the lands they inhabit, it has been pursued with vigour all over Africa. This has either been as a result of an overt national policy or administrative action, as in Tanzania's 'villagisation' programme, or in response to a crisis of dependency as was the case in the Sahel after the 1970s and 1980s droughts, or as the inevitable consequence of land tenure reform and the push for 'privatisation' sponsored by some Western aid donors.

Whatever the potential merit of providing improved welfare services to rural populations, the settlement of pastoralists has profound implications for common property resources management. This is why pastoralists like the Karimojong in Uganda have so consistently and effectively resisted permanent settlement, often only conceding it when poverty provides no other choice.

Demarcation of communal lands and their allocation to villages has the potential to disrupt customary land use patterns. Village boundaries not only divide common land into discrete administrative units, but also may deny access to resources that were otherwise accessible to communal land users outside the village who traditionally had access to wider ecological land use units. This access is critical where resources are not equally distributed or in times of drought when movement to areas that retain resources outside a home range is necessary to maintain herd productivity and thereby ensure people's survival.

Formal administrative frameworks tend to confine land management to village authorities with jurisdiction restricted to the area within village boundaries, and provide little or no scope for land use management over a wider area. Initially, herders may retain access to resources in a number of villages on an ad hoc basis or because boundaries are ill-defined. However, this provides no security in the long term as local government, even at village level, cannot necessarily be relied on to maintain communal access to pastures, nor does it have the mandate or capacity to facilitate inter-village land use planning (Lane 1996).

Titling and Land Use Planning

Land titling is often an integral part of land reform measures. The belief behind this is that only through registered title can landholders have a sufficient level of security to invest in land for increased productivity, encourage them to make improvements, protect resources from over-exploitation, and induce lenders to finance improvements through the provision of credit.

Titling presents a double-edged sword to pastoralists who, while recognising many of the obvious advantages of having registered title as protection from land grabbing, find it poses problems for maintaining extensive seasonal land use patterns over a large area. A detailed study of a pastoral village in Tanzania shows that those Barabaig herders who settled did so out of poverty and were forced to compromise their herding strategies by limiting the extent of their migration to the distance their herds could travel to and from the homestead in one day. The concentration of animals within the village has had an adverse ecological impact, encouraging a trend towards agro-pastoralism and a further decline in levels of production (Kjaerby 1979).

Formal land use planning often accompanies settlement and complicates matters for pastoralists because of the inability of planners to reflect the complexity and necessary flexibility of customary land tenure arrangements which permit mobility (Behnke and Scoones 1992). According to a rapid rural appraisal in a pastoral area of Tanzania by the World Bank the planning process was neither transparent nor intelligible to villagers. Little account was taken of villager perspectives, nor detailed environmental knowledge attained beforehand. Hence the plan is likely to damage the local pastoral economy, degrade the environment and prompt the out-migration of pastoralists from the area (Johansson 1991).

Sahelian pastoralists have experienced similar difficulties. However, two new initiatives in francophone Africa may improve their prospects.

The 'Gestion de Terrior' (gestion = management, terrior = village lands) approach in a number of West African countries arose out of the recognition that there was a lack of clarity regarding land rights, that development performance was poor, that the ability of the state to administer lands was limited, and that natural resources were subject to adverse trends. 'Gestion de Terrior' operates within the three interrelated systems of technical interventions, socioeconomic institutions, and legal rights and responsibilities. It aims to tackle these problems by clarifying rules of tenure, redefining local community rights and facilitating participatory diagnosis of problems (Toulmin 1994).

Where this approach differs from what is happening in East Africa is that it is based on territories that more closely reflect ecological land use units, and enables local land users to participate more fully in the planning process through problem identification, design and implementation of management plans, election of local resource management committees, and monitoring and evaluation of land management. 'Gestion de Terrior' reflects a fundamental shift in relations between local land users and the state by redefining local community responsibilities and rights in relation to land. However, it does not fully overcome the problems of access to resources beyond territorial boundaries, nor does it fully protect pastoral resources from alienation, since land management committees are usually dominated by non-pastoral interests.

The government of Niger is trying to address the problems of tenure security by drafting a rural code ('Code Rurale') that attempts to clarify the profusion of contradictory laws, decrees, ordinances and circulars that currently constitute the country's land law, and elevate customary land tenure rights to formally recognised laws. The drafting process involves public hearings, the commissioning of scientific studies and the encouragement of widespread political debate (Lund 1993). While the challenge of formalising customary laws remains, the high level of involvement by local resource users offers the possibility of protecting the rights of different user groups.

With the certain knowledge that land tenure is to be reformed, and in the time it has taken to draw up the code, rural landholders have moved quickly to strengthen their claims to land. This has exacerbated conflicts, particularly between farmers and herders, as those with cultivatable land have become less interested in allowing herder access in case this weakens their claim. The attempt to base the code on customary land tenure has also led to an institutional quagmire in defining what makes up customary law in any given location, and poses a number of very important questions that need answers. Is it only old rules, or contemporary rules interpreted by a traditional leader? Does that leader represent the best interests of the whole community? What traditional institutions have authority to attest to such rules? How can consideration of equity be attained through traditional and sometimes undemocratic institutions? What relationship will traditional institutions have with wider legal and administrative state structures?

Even though the answers to these questions have yet to be found, the process continues and a return to the situation before the code is now impossible even if it were desirable. There are, however, very good reasons why the process should be carried on. It is likely that only through the involvement of local people in the formulation of land law, with reference to those customary provisions with which they are most familiar, can land tenure apply to the diverse environments and myriad of land tenure arrangements that are found. Its achievements will hold important lessons for governments which face the challenge of resolving conflicts and providing greater land tenure security throughout rangeland areas of Africa.

Privatisation

Privatisation of communal rangeland is based on the premise that only through such measures can degradation be arrested and productivity be improved. However, there is no certain correlation between individual title and higher levels of productivity. A growing number of case studies (some of which are found in this book) also show that privatisation of pastoral resources has resulted in less effective rangeland management.

Under a 'new' national tribal grazing lands policy in Botswana, for instance, communal grazing lands are to be fenced. When this happens there are good reasons to believe that this will only effectively allocate grazing land as de facto private property to an elite of wealthy land owners who can acquire exclusive use of the best grazing areas by making the necessary high investment in boreholes (White 1992). The resulting skewed access to rangeland associated with improved availability of water enables individuals to accumulate an increasing share of the national herd at the expense of poorer herders who cannot afford to invest in drilling for water. On the other hand Abel and Blaikie (1990) have shown that a 'tracking' strategy on non-enclosed rangelands will allow higher numbers of animals to be kept on the range by utilising 'surplus' forage in wetter years.

Experience with sub-division of group ranches in Kenya also highlights the problems of privatising communal grazing areas. Apart from the problem of denying communal access to resources, it has allowed some private landholders to acquire additional plots for themselves, and others to sell land out of the ranch thus fracturing the commons into potentially unsustainable individual grazing units. Private landholders are able to overcome this problem by 'dual grazing' their stock on what remains of communal lands when forage on their plots is exhausted, or when they want to regenerate their own pastures. In this way they are conserving their own pastures at the expense of those who rely on communal lands (Galaty 1980, Oxby 1981).

The fundamental premise of the argument for privatisation that it will arrest degradation of resources and increase productivity of rangeland resources has been shown to be flawed. While research in Thailand lends support to property rights theory by concluding that 'security of land ownership in Thailand has a substantial [positive] effect on the agricultural performance of *farmers*', particularly for gaining credit for investment in improvements for greater productivity (Feder et al 1988), research in Africa suggests that a direct correlation between individual title and higher levels of production is more elusive. In a comprehensive study of household survey data from Ghana, Kenya and Rwanda in 1987, Place and Hazell found with few exceptions that 'land rights are not found to be a significant factor in determining levels of investment in land improvements, use of inputs, access to credit, or the productivity of land' (1993). Although these studies did not include pastoral communities using extensive range resources with communal tenure arrangements, they nonetheless undermine the basis for the ambitious registration and titling programmes that are underway throughout Africa.

Settlement of nomads, land titling, formal land use planning and privatisation in the dryland pastoral sector of Africa have clearly failed to meet objectives, and in doing so they illustrate the weakness of approaches based on inappropriate theories. The examples illustrated in this book will show that land tenure reforms advocated by those who adhere to the 'tragedy of the commons' argument provide little protection for pastoralists from alienation of their lands, limit strategies for herd movement and making the most efficient use of resources, deny them the means to cope with risk and uncertainty, take little account of resource diversity and the tenure systems that manage them, and marginalise poorer members of pastoral society (Lane and Moorehead 1994).

Policies for privatisation, titling, settlement, and the nationalisation of land have all clearly failed to achieve the goals they propose. In doing so they expose the weakness of the 'tragedy of the commons' as a viable approach for rangeland development. Adherence to this approach shows that it:

1. provides pastoralists with little protection from land alienation;
2. can lead to the incompatible double allocation of rights to pastoral resources;
3. limits mobility and scope for the adoption of risk-averting strategies;
4. takes little account either of resource variability, the complexities of customary land tenure arrangements, or the diversity of herders, often resulting in the marginalisation of weaker and poorer social groups, particularly women;
5. diminishes access to key resources that sustain the wider pastoral ecology.

Further, the costs of implementing private property systems are huge in terms of the resources and time required for survey, registration and dispute resolution.

Four observations can be drawn from this brief overview of the theories and major approaches to land tenure reform:

First, centralised, uniform and imposed land tenure reform is not likely to succeed in enabling pastoral land users to produce more and protect their lands from over-use. On the contrary, it is likely to heighten insecurity, undermining productive potential and putting land at risk of degradation—the result that is hoped to be avoided. Experience shows that formal structures devised by central authorities do not accommodate adequately the necessary complexities and flexibility of customary tenure arrangements, and in the chaos created by their imposition opportunities exist for outsiders to take up pastoral lands to the exclusion of the original pastoral inhabitants.

Second, privatisation of land provides neither equity nor efficiency for pastoralists in non-equilibrium environments. Policies of nationalisation and privatisation can have debilitating effects on communal tenure systems, without providing an effective, alternative regime.

Third, there is clearly a need to move away from technical solutions and to take greater account of social, economic and political considerations—away from simply trying to improve productivity on private land towards improving the manner in which reciprocal tenure arrangements can build consensus between resource users as stakeholders in the management of rangelands they depend on for their livelihoods.

Fourth, increased attention needs to be paid to the physical characteristics of resources and their relationship to tenure systems, and on the relationship between tenure systems and institution building. It is being increasingly appreciated that the brokerage of interests between different interest groups (between and within production systems) is a strategic issue for creating more equitable and efficient land use.

CASE-STUDY REVIEW

The overviews reveal important trends in land tenure in the African pastoral sector.

There are moves afoot in a number of African countries to review land policy and address the issues raised above. For example, in 1991 the president of Tanzania instigated a land commission to look into all matters related to land. Concurrently, the Ministry of Natural Resources, Housing and Urban Development is overseeing the formulation of a new national land policy that will be supported by a single legislative instrument, Basic Land Law, that will bring together as one the myriad of legislative instruments and executive orders dating from colonial times. However, while reappraisal provides the opportunity for reform it is by no means certain that new policy and legislative instruments will resolve the problems of land tenure insecurity for pastoralists. This is made clear from analyses in the following case studies.

The Colonial Legacy

Many of the difficulties faced by pastoralists today have their origins in measures taken in colonial times that had almost universal adverse impacts on pastoralism. Under colonial rule, pastoral land was often assumed to be unowned or underutilised. Land use was conceptualised mainly in terms of its potential for cultivation agriculture, particularly cash crops. As a result, due to the particular nature of pastoral land use patterns based on communal land tenure, traditional lands were vulnerable to settler occupation and alienation by the state for commercial production and wildlife conservation.

In Uganda, for example, the Crown Lands Ordinance of 1903 radically transformed land tenure by giving the British colonial authorities power to alienate customary lands and issue freehold and leasehold titles. The overall tendency was to support the privatisation of land ownership, get pastoralists to settle, and encourage cultivation rather than extensive cattle keeping. Colonial treaties with Maasai leaders in Kenya between 1904 and 1913 led to over 50 per cent of their traditional lands being surrendered to white settlers. Large tracts of remaining pastures were occupied as Crown Land on the grounds that these were not being fully utilised. In Sudan, the British Land Settlement and Registration Ordinance of 1925 recognised customary tribal ownership of communal land. However, the customary communal land tenure systems only operated in 'homeland' areas and denied rights in areas further away, such as where pastoralists moved in times of drought.

In addition vast tracts of pasture land were acquired for wildlife conservation. Ironically, wild animals are free to wander from parks into pastoral areas, but pastoralists are denied access to wildlife reserves. The Royal National Parks Ordinance of 1945 established game parks in Kenya, and denied pastoralists rights to resources within park boundaries and game reserves, in which pastoralists retained only restricted rights of access. The National Park Act of 1952 was designed to protect wildlife to the exclusion of pastoralists in Uganda, who had traditionally shared resources with wildlife without adverse impact. Also, the whole of Karamoja, covering nearly 7000 square kilometres (km^2) was declared either a controlled hunting area or game reserve. The adverse effects of denying access to an important permanent water source with the creation of Lake Mburu National Park were mitigated by de facto access until only recently, when resident Bahima herders were evicted.

From the case studies it seems that there was little difference between the approach to land adopted by British and French colonialists. In both anglophone and francophone Africa the state alienated land from customary rights holders to further colonial interests. For example, the French landholding code introduced in Senegal in 1830 enabled the colonial administration to take over traditional lands said to be 'vacant and ownerless'.

Within the inland delta of the river Niger in Mali, the pre-colonial Fulani state tried to formalise a complex system by which different land users gained access to resources at different times under the control of a traditional management structure. The French colonial administration codified the nineteenth century Dina system and centralised control over the many and varied resources of the area that were managed by a complex system of rights and regulations. Control over what were termed 'vacant lands' was achieved through legislation, declaring all such lands as the property of the state. In doing so the state created two parallel forms of land tenure: customary, governing areas under continuous production (as defined by the French) and managed by customary authorities, and statutory, covering vacant lands to which private title could be acquired through legislative procedures. The administration tried to coordinate management of natural resources and fix the hitherto flexible times of access to the various pasture areas of the delta. This has undermined traditional local authorities' control over access, whose decisions were based on detailed assessment of range conditions, and opened up pastures before they were ready to greater numbers of livestock herders.

Before nationalisation, entry to Niger floodpastures was determined by a hierarchy of predetermined rights of access, and an assessment of pasture quality by traditional community leaders. Today dates of entry are fixed by central government on a calender basis, and any citizen of the country has rights of access. This has created structural chaos by undermining the ability of local land users to manage resources in their own interests and control levels of use, while at the same time failing to provide an equitable and effective alternative to the customary land tenure system it replaced.

Legislation

Legislative instruments have been used by most independent African states to legitimise alienation of pastoral lands. The Sudanese government, for example, has passed various laws culminating in the Unregistered Land Act of 1970, which effectively alienated large areas from pastoralist control. As it is, extensive communal land ownership is now only found in the south of the country, but its future is uncertain as Dinka, Nuer, Shilluck, Mandari and Murle pastoralists have been ravaged by the civil war in recent years. Pastoralists elsewhere in Sudan have been losing land to agricultural schemes under leasehold tenure or by virtue of title through private ownership. The Sudanese government has focused on grain production and where water schemes have been provided for pastoralists, these have led to land degradation since access to them has not been effectively controlled. Pastoralists in Sudan have also lost land because of the general weakening of local authority and their ability to serve their own interests. Pastoral interests have not been served by either the Local Government Act of 1971 or the Regional Government Act of 1981, the latter of which abolished native administrations which had previously operated to defend pastoral interests.

In Senegal, Law Number 72—25 of 1972 ceded authority for rural areas to local authorities and the administration of grazing areas to local communities, working through rural councils. However, the potential advantage of having local land use authorities has been undermined because zoning into different economic activity areas prevents the development of pastoralism on its own terms. State plans for economic development do not recognise the potential of pastoralism, unless it is transformed and integrated with agriculture and unless pastoralists become settled. Large tracts of pastoral land and water sources have been alienated for agriculture and forestry projects from the 1960s, or for conservation as with the Niokolo Koba National Park in 1954. Pastoralists have been excluded from the provisions of Law Number 72—1288 in relation to the allocation of land for productive use and are also regulated by the provisions of Decree Number 80—14 applying to water points and livestock tracks. The settlement of pastoralists has also been imposed by legislative authority. Transhumance for Peuhl pastoralists in Senegal has been severely curtailed by legislation nationalising land (Law Number 64—46) and designating specific areas for livestock keeping (Decree Number 64—573).

Although pastoralists are severely regulated in this way in Senegal, this has led to more thought being given to the specific needs of pastoralists in relation to access to rangeland resources. However, as more and more pasture land is lost to agriculture, forest reserves and national parks, and as livestock numbers increase, pressures on pastoralists have intensified as mobility has become more and more constrained. This has led to outbreaks of violent conflict between pastoralists and agriculturalists, and was a factor in hostilities between Senegal and Mauritania in 1989.

In Mauritania, Decree Number 83.127 nationalised land and abolished customary land tenure in favour of state and private ownership. Despite this, however, traditional communal tenure systems are maintained in practice, albeit with increasing difficulty.

Similarly in Uganda pastoralism based on the utilisation of communal lands is regarded as unproductive. The 1975 Land Reform Decree declared all land to be public land vested in the Uganda Land Commission. This has had varying impact in different parts of the country. Communal grazing among the pastoral Bahima in the southwest has largely been phased out by private ranching schemes, while in Karamoja private ownership of land is being resisted. Although livestock development policies remain biased towards ranch development, the Ranch Restructuring Board has tried to accommodate landless Bahima pastoralists on ranches which are considered excessively large and undeveloped. Proposed land law reform recommends formal registration of communal grazing land in the names of groups of rangeland users.

Legislation in Kenya has attempted to formalise traditional communal forms of land tenure. The Group Representatives Act of 1968 provided for the transformation of traditional communal tenure into group ranches. However, since 1982 one of the government's directives has insisted on the sub-division of 'group ranches' into individual holdings leading to the disintegration of Kenya's pastoral commons and the acquisition of private titles and the sale of lands to non-residents (Galaty 1994).

A similar approach was attempted in Tanzania's Maasailand with the 1964 Range Development Act. The Maasai Range Commission was charged with the task of allocating land to ranching associations in the Range Development Area. All customary rights to land were extinguished with the issuing of group title. By 1980 only eight out of a planned 22 associations had been formally registered, and it was eventually superseded by the Villages and Ujamaa Villages Act of 1975 which required such ranches to be re-registered as villages, which in turn has constrained pastoral land use and undermined pastoralists' welfare.

Donors, Differentiation and Degradation

International donors have been involved in the reform of land tenure and the shift away from communal and customary to more private and statutory forms of land tenure. This transformation has led to the displacement of pastoralists through alienation of traditional pastures for private and state farms.

For example, in Tanzania the Canadian International Development Agency (CIDA) supported a large scale government wheat scheme that alienated over 100,000 acres of the Barabaig pastoralists' most productive pastures (Lane 1996). In Kenya, Turkana tree tenure rights have been ignored in the development of the NORAD (the Norwegian Agency for Development Cooperation) and FAO (the UN Food and Agriculture Organization) funded Turkwell Dam Project. USAID (the United States Agency for International Development) has long been involved in support for private ranching schemes for commercial beef production in Uganda. And in the past non-governmental organisations (NGOs) such as Oxfam have been active in Karamoja supporting projects which focused on cultivation and settlement schemes which tended to weaken pastoral systems.

The failure of donors to support pastoralism effectively and the active undermining of pastoral land tenure arrangements have resulted in greater social and economic differentiation among pastoralists. This is seen most critically in terms of access to resources where a growing number of poorer households are being denied the means for survival, and traditional inheritance and redistribution mechanisms are beginning to break down making it increasingly difficult for people to cope. This has had particularly adverse effect on women who have not been accorded equal rights with men to livestock or land.

The Mauritania case shows that after the 1973 drought a new class of commercial livestock producers, drawn mainly from the ranks of traders and government officials, acquired control of an estimated 40 per cent of the country's livestock. Very little investment in pastoral production is now made by government and assistance is offered to pastoralists only if they agree to settle or form pastoral associations, which have been established since 1990 and which grant herders usufructuary rights to land.

Impoverishment of pastoralists has led to an increase in the cultivation of crops on pasture land and the development of agro-pastoral economies as pastoralists have attempted to attain greater food security. The combined effects of land lost to cultivation and a reduction in the area left to pasture has caused degradation. In Tanzania, the growing scale of farming, together with the expansion of areas set aside for wildlife conservation and uncontrolled charcoal production in traditional pastoral areas, has resulted in the degradation of those pastures that remain available for herding. In Kenya, environmental degradation is associated most with government-provided wells as a result of failing to take account of customary land tenure arrangements, and effectively controlling who has and who does not have access to water and therefore the surrounding pastures. Adverse environmental impacts are also caused by overt actions as well as omissions. In Mauritania, for example, destruction of the rangeland environment is encouraged by the commercial patronage of the charcoal business by senior government officials.

Power Relations

Struggles involving pastoralists seem to be increasingly fought along class lines. Despite state legislature, the real battle over pastoral land seems to be taking place between those pastoralists interested in preserving communal land tenure and a class of absentee livestock owners wanting open access to pastoral resources. In Senegal, pastoralists are said to be mainly illiterate and powerless in relation to bureaucratic and commercial interests. To date attempts by pastoralists to organise themselves in West Africa have not been very successful.

In Sudan, the Department of Range and Pasture Management is entrusted with the task of looking after pastoral interests, but is unable to give pastoralists sufficiently high status to influence national development plans. As a result, pastoralists are increasingly faced with destitution as they are no longer able to retain control over resources. Few pastoralists are represented by organisations that can negotiate their interests. To date they tend to rely on sympathetic international organisations to lobby on their behalf. However, developments in Tanzania with the advent of a growing number of indigenous pastoral NGOs might advance prospects for pastoralists. Publicity about land alienation and violations of rights in Tanzania prompted the president to respond with the appointment of a commission of enquiry into land matters. It remains to be seen whether its recommendations in support of pastoral rights to land will be accepted, and how long widespread abrogation of pastoral land rights will be allowed to persist.

VIEWS ON THE WAY FORWARD

Rapid population growth in Africa, its spill over onto rangelands, and the ever growing area of land being cultivated for crops is making it increasingly difficult for herders to maintain their mobility. Of particular importance is the take-over of key wetland pastures by farmers as private agricultural property (both de facto and de jure), removing from herders' control resources they rely on to sustain production from more marginal resources at other times of the year (Scoones 1991). Herders can only make use of marginal grazing areas because they have access to wetlands during the dry season. Loss of access to the latter is likely to bring added pressures on land use, constrained productivity and adverse impacts on the environment (Lane and Scoones 1993). Thus there is an urgent need for tenure systems that permit mobility and assure access to key resources.

Several important political and economic processes are currently underway in many countries of Africa. They bring with them the decentralisation of authority over natural resources, liberalisation of political activity, structural adjustment and the conditionality of aid, each presenting both opportunities and threats to pastoral groups. Opportunities include the growing acceptance of participatory processes in development action in which local land users can choose their own priorities and assume responsibility for management of natural resources. However, threats from the same processes include the risk that decentralisation may be coopted by non-pastoral interests which are better represented in power structures, causing herders to miss out on the opportunities offered by land registration and planning initiatives.

New thinking on land use in non-equilibrium areas places rangeland resources at centre stage in the debate on appropriate tenure regimes for pastoral production in Africa (Behnke and Scoones 1992). As pastoral resources are often subject to high variability within and between seasons and across large areas, future tenure systems should support herders' tried and tested strategies of mobility. The 'new directions in African range management policy' proposed by Behnke (1992) have very specific tenure implications:

1. a rangeland's carrying capacity has to take into account the management objectives of rangeland users as well as the vegetation characteristics found on the range, implying devolution of authority for management to local herders rather than the imposition of centralised control;
2. tenure systems must allow herders to move at short notice so as to capitalise on areas of high productivity, and by extension not be incumbered by time consuming procedures;
3. tenure systems must provide secure access to a range of ecological zones.

Tenure systems that embody these attributes are necessarily communal. On Africa's rangelands they belong to cohesive groups of land users which make up pastoral society, which share reciprocal access agreements with neighbouring groups who have the same interests in and dependency on a defined set of resources. Within each larger grouping preferential access and authority over management of a given resource are devolved to smaller social units (that is, clans or lineages) with lesser rights accorded to groups with whom more distant social relations are maintained, while outsiders may be entirely excluded. Legitimation for such an order may be based on historical factors including the longest claim of occupation or a predominant claim acquired by use of force.

Pastoral systems have evolved from the physical characteristics of the resources that make up the range, allowing a quick response to unpredictable natural events so as to maximise access to available pasture, while providing more regular access to a set of disparate resources over a number of seasons. It achieved this by vesting 'ownership' of resources in larger social groupings that provide the control necessary to retain occupation of the range, while at the same time providing a simple and quick decision-making process through kinship links, legitimised by a shared culture, and providing a set of clear rules understood and accepted by everyone.

However, contrary to the notion of bounded property inherent in common property resource theory, in which communal land users are confined to a well-defined range area, Behnke (1994) argues that the inherent dynamics of non-equilibrium environments imply a greater degree of flexibility than can be attained through demarcated group territories and formal management institutions. Range boundaries are more of a political nature based on 'self-interested—and contradictory—claims to resources by different parties'.

Accommodation of the traditional land use patterns of pastoral groups does not mean the adoption of a new orthodoxy to replace the old. There is no single panacea to the problem of land for pastoralists. Each location has its own specific context that must be taken into account before remedial measures are imposed. The cases from Mali and Mauritania illustrate well the complexity that pertains in some pastoral areas. Despite the less complicated pictures painted in countries like Sudan and Uganda, it can be expected that issues of power and tension in relations between different resource user groups, including cultivators and fishermen, are likely to prevail unless measures are taken that work towards an assurance of sustainable management of resources with consideration given to issues of equity and justice.

It is perhaps only wishful thinking to hope that traditional land tenure systems could provide coherent management of rangeland resources today. It may be too late to re-animate customary communal land management when laws and administrative provisions about them are changing in the opposite direction. There is evidence to suggest that kinship and other social linkages that once held pastoral land tenure systems together have been either destroyed or severely undermined. Diverse interests within pastoral society as well as wider economic and political influences are highlighting the increased divergence between rich and poor herders. Any pastoral land tenure policy that ignores these divisions is therefore unlikely to succeed.

In the light of the challenges posed by pastoral groups in non-equilibrium environments, Behnke (1994) puts forward six organisational principles with which to devise new rangeland management structures:

1. The administration of resource use should at least pay for itself. For management to be sustainable, costs should not exceed productivity.
2. Management systems must cater for the inherent characteristics of range environments (sometimes massive physical size of management units, disaffected and often belligerent pastoral communities, and very great environmental variability).
3. Legal rights to control access to and management of resources must be affirmed.
4. The management approach should shift from regulation to allocating and upholding rights of access to resources. The role of central authority becomes one of arbitration in cases of conflict rather than direct intervention.
5. The use of directives (executive orders and so on) and legislation to dictate the content of property rights should be avoided. Customary land tenure arrangements are too complex to codify. The state should provide a framework within which conflicting claims for rights of access to resources can be

assessed and resolved.
6. Range management should be focused on 'key' or limiting resources — those most crucial to productivity of the wider dryland ecology, or the most regular subject of conflict. 'Focal point management would concentrate on those resources which lay at the heart of the [dryland] production system, and devote much less effort to the clarification of rights to resources which were very abundant, of low or erratic productivity, or geographically extensive and difficult to police' (Behnke 1994).

There are, however, legitimate doubts over whether policies that take account of these principles can address adequately issues of equity within the pastoral sector. There are many examples of key resources being taken over by wealthier and more powerful groups within pastoral societies, and support for traditional pastoral structures might only help these elements to achieve greater control over the most valued resources. For example, within agro-pastoral areas, herding groups using dispersed pastures lying between farming areas (such as the WoDaabe in Niger) might be marginalised in this process (Thebauld 1993). Registration of titles to village land in Tanzania can provide protection from alienation of pastures to outsiders where village councils represent pastoral interests. However, where they do not or when enticed they can use their authority to allocate land to whom they please by corruption of land allocating procedures (Lane 1993a). Finally, it is by no means clear that all pastoral societies have the capacity to organise themselves sufficiently well to ensure sustainable land use under current conditions.

In the face of a conflict of interests pastoralists have resisted alienation from their lands. They are beginning to organise themselves around the issue of land rights as indicated by the burgeoning number of pastoral NGOs in East Africa (Galaty et al 1994). Very often this has pitted pastoralists against the state, its surrogates and supporters. While confrontation has brought some benefits in terms of publicity and even legal cases won, it has also had negative impacts. By antagonising officialdom, pastoral organisations have been restricted in their activities, public information has been suppressed, and some protagonists have been interned. If this is to be avoided in the future, different and more diverse strategies will be required that are based on a mixture of measured confrontation with more conciliatory approaches involving such activities as lobbying and consensus-building between parties to conflict.

When considering pastoralists' capacity to organise themselves there is a tendency to assume that pastoral society is homogeneous and all members are equally affected by the changes going on around them, particularly those related to land tenure. Research too often fails to distinguish between different interest groups and classes of people within society. Very often elites within the community may play as large a role in altering tenure arrangements as outside forces. The impact of tenure changes on pastoral women is not yet fully understood, nor is their role in managing natural resources. As such they present an important focus for future research to inform action.

A challenge remains for donors and governments to collaborate in support of a new approach for the development of Africa's rangelands. This will require the redefining of the role of the state vis-à-vis pastoral communities. Whereas before the state was seen as all powerful, now many African states have found that they are losing control of the margins, becoming impoverished by economic mismanagement and beleaguered by social and political unrest. This has created a vacuum of power over how natural resources are controlled at the local level. Until most recently this vacuum was being exploited without restraint by interests that had no stake in the persistence of pastoralism nor concern for the long term environmental consequences of their actions.

Pastoralists have had to endure administrative chaos and the dispossession of their lands for over 50 years. Now they are beginning to organise around the issue of land rights in the hope they can salvage what remains. Pastoral organisations are beginning to challenge the state through litigation and use publicity to apply pressure for change. This is being recognised as a just cause by international human rights organisations and donors are beginning to place conditions on aid flows to control the behaviour of the state in the name of 'good governance' consistent with Agenda 21 of the United Nations Conference on Environment and Development. If pastoralists are to maintain their livelihoods on land that remains to them or on what they may reclaim and thereby become again custodians of the commons for future generations, then they need above all else to enjoy security of tenure. Even though pastoral organisations are under-resourced and inexperienced in the task of representing their interests and addressing the problems of natural resource management, they offer the best hope of doing so sustainably if they are able to acquire the necessary capacity to assume a more formative role in the development of conducive policies, sympathetic laws, ecologically sound management, and efficient administration of lands, and are supported in this by governments, development agencies, and donors alike.

2

KENYA

Mukhisa Kituyi
African Centre for Technology Studies

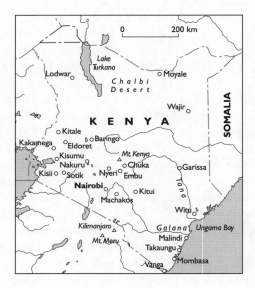

Map 2 Kenya

INTRODUCTION

About 83 per cent of the land area of Kenya (Map 2) is categorised as arid or semi-arid. This is land with less than 600 millimetres (mm) of annual rainfall, and incapable of sustaining traditional rain-fed agriculture. This area is the traditional habitat of the pastoral and agro-pastoral peoples who account for about 15 per cent of

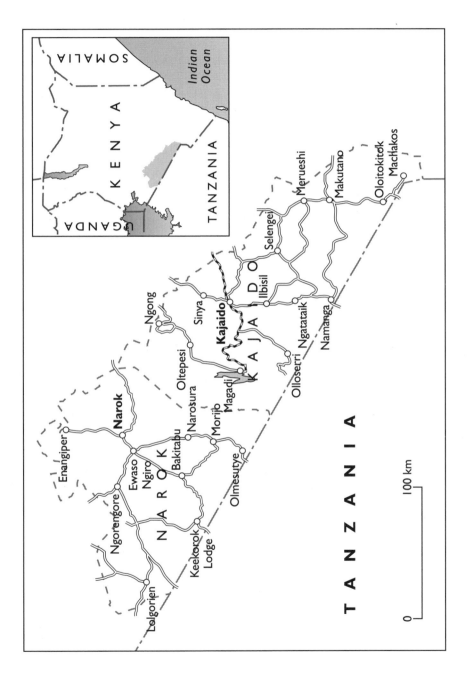

Map 3 The Two Maasai Districts of Kenya

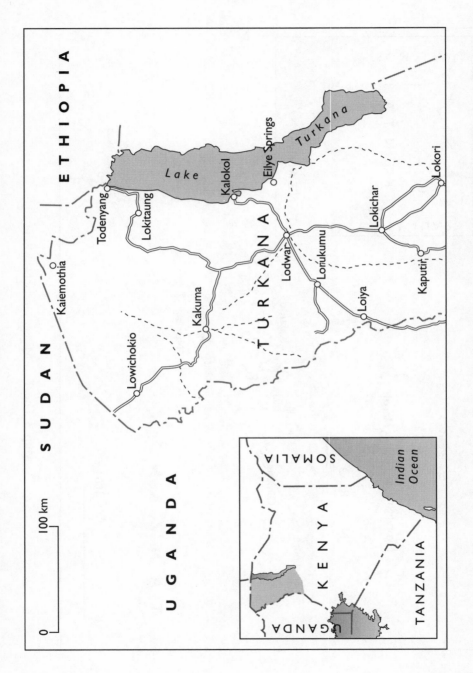

Map 4 The Turkana District of Kenya

the country's 24 million inhabitants. Of all Kenyan pastoral peoples, the Maasai have experienced the most extensive changes in land tenure and, as such, represent an important pointer of trends for other Kenyan pastoral societies.

Kenya's pastoral population can be divided into three main groups. The southern group comprises the Maasai and their closely related Samburu cousins. They occupy the crescent stretching from Samburu and Laikipia districts west of Mount Kenya, and extending south and southeast to Narok and Kajiado districts bordering Tanzania (Map 3). These are members of the Ol-Maa-speaking peoples, who for centuries dominated central Kenya and northern Tanzania. This cluster of pastoralists is almost completely surrounded by agricultural communities, and borders major urban centres like Nairobi. The land is the most heavily populated among the rangelands. These factors play a critical role in the land question. The group has also experienced relatively substantial development of communications infrastructure and has increasingly become integrated into the wider economy.

The northern group occupies the whole northern half of the country, stretching from Lake Turkana to the west, to the Somali border in the east, and from the middle of the country to the Ethiopian border. This group comprises members of Cushitic and Oromo speakers who have spread into Kenya from the near north. They include Somali, Gabra, Merille, Boran, Rendile and Tana River Oromo. Theirs is the driest land in Kenya, characterised by ecological fragility, low population densities, limited threat from agricultural neighbours, and a poor history of state investment in infrastructure and market development.

The 300,000 Turkana occupy the northwestern corner of the country bordering Uganda, Sudan, Ethiopia and Lake Turkana (Map 4). Culturally, the Turkana belong to a separate group comprising the Karimojongh of northeastern Uganda, the Iteso of eastern Uganda, the Toposa of southern Sudan, and the Dodoth of Ethiopia and Sudan. While sharing some of the ecological features of the other two land groups, the area occupied by the Turkana was a predominantly closed district to the rest of Kenya until the past two decades.

POLICY CONTEXT

Official thinking and practice on land tenure in Kenya's rangelands has been characterised by responses to three major challenges:

- the rights of local pastoral peoples in the face of competing interests;
- the relationship between pastoral land use and wildlife conservation;
- regulation of tenure which seems to either reduce depletion of resources or increase their exploitation.

PASTORALISTS AND THE STATE

Land rights became a problem in the colonial period: early colonial settlers were allowed to claim exclusive rights over lands under pastoral control. These claims were made in two ways.

The first was the use of treaties by which locals surrendered their rights to some land. This method, which entails an acknowledgement of local rights over the land, was first used on land occupied by the Maasai and on a ten mile coastal strip which belonged to the Sultan of Zanzibar.

Between 1904 and 1913 the colonial government signed treaties with what they regarded as Maasai chiefs, by which the Maasai surrendered over 50 per cent of their land to white settlers. These treaties were accompanied by an official undertaking that all the remaining Maasai lands would be held in trust for their exclusive use. This undertaking, however, did not prevent subsequent land appropriation.

The second method was more forceful and widespread. This was the proclamation of what the government considered unoccupied lands as Crown land. The main proponents of early colonial land alienation argued that Africans did not own land, rather they enjoyed only immediate use rights in occupied lands. Thus the new sovereign authority could easily take on lands rarely used by the Africans. As one of the spokesmen for the British settlers in Kenya put it at the start of the century:

Africans owned land only in terms of occupational rights, [and thus] unoccupied land reverted to the territorial sovereign. [But since] small chiefs and elders [were] practically savages in whom sovereignty could not possibly reside, the only reasonable alternative was Her Majesty's Government. (Okoth-Ogendo 1991)

It is widely assumed that this line of thinking has been discredited, but it has had a significant impact on latter day land policy in Kenya. The assumption that land not immediately claimed by an individual is unowned has been behind land acquisition for game sanctuaries and forests, on both sides of independence. It also supports the state's claim that better usage justifies acquisition.

The scarcity of agricultural land in Kenya, in the face of an increasing population and limited salaried employment, has encouraged the state to support substantial rural-to-rural migration. This has been a major source of insecurity among pastoralists (Leys 1977, Kitching 1980, Juma 1989, Swynerton 1955). One minister for lands proclaimed that official policy means that there is no such thing as tribal land in Kenya, and the government can settle anybody anywhere in the country (Weekly Review, 30 March 1984).

The results of this policy have been varied. For the southern pastoralists, the Maasai and their Samburu cousins, the proximity to agriculturally based neighbours with a history of intermigration, the abundance of cultivable land, and the collapse of internal military power, might have facilitated peasant infiltration of the pastoralists' land before independence. Since independence, however, the government has done nothing to stop the massive immigration of entrepreneurs and displaced cultivators into pastoralists' land (Kituyi 1990).

In northern Kenya, harsh ecological conditions, the distance from cultivators, attacks by Somali bandits and the treatment of these lands as a closed frontier district all combined— before and after independence—to restrain both peasant infiltration and the development of a coherent tenurial policy.

PASTORAL LAND AND GAME CONSERVATION

The abundance of game species in Kenya has also played its part. The lobby for game conservation in Kenya has gone through three phases. Before 1945 game was protected on an ad hoc basis in tandem with the systematic extinction of traditional hunting ('poaching'). Between 1945 and the late 1970s protected game parks were established; since then 'mining the parks through tourism' has occurred (Berger 1989).

In the colonial period, the rangelands were the focus of policy debate on the creation of game sanctuaries. This debate led to the Royal National Parks Ordinance of 1945. This established exclusive game parks to which pastoralists surrendered their rights, and national game reserves in which limited pasture rights could be exercised on condition that wildlife would not be disturbed. The first two national parks (Nairobi in 1946 and Tsavo in 1948) and the first national game reserve (Amboseli in 1948) were established on, or adjacent to, Maasai land.

The Wildlife (Conservation and Management) Act (Cap. 376) of 1976 further emphasised the sanctity of game protection and banned game hunting. This removed the right of pastoralists to kill game for food. Although this act mentions 'utilisation' by the local community as part of conservation, no clear provision is made for this.

Over the years, such world renowned game sanctuaries as Maasai Mara, Samburu and Marsabit have been established, depriving the pastoralists of some of their most important land in the process.

POLICY CHANGES

In Kenya, the colonial authorities did not develop a consistent land policy for

Africans. This changed after the war when orderly transfer of power to an independent government became a necessity. During the state of emergency in the 1950s, the Swynerton Plan (1955) pointed out the need to create an African agricultural middle class that could protect the rights of private property at the time of independence.

Although all state legislation has acknowledged communal rights to land, this right has been seen as just one step in the transition to other forms of private property regulation. Among cultivators, traditional communal rights (mostly inherited) have been the basis of claims in the process of sub-division. Among pastoralists, traditional rights have been claimed as the basis for membership of a limited group with exclusive rights to land, or for individual ownership of sub-divided communal lands.

In Kenya's rangelands, substantial variation between law and practice occurs. The most comprehensive policy statement is the Group Representatives Act of 1968. This provided for the transformation of traditional communal tenure to a more private tenure system. Meant primarily for the southern pastoralists, the act provided for local groups to be assigned a section of land registered as 'group ranches'.

Groups of pastoralists in the Narok and Kajiado districts were associated with group ranches, each having a ranch management committee seen as representing the interests of all those with customary user rights to the land. Membership— and the exclusion of outsiders—would be determined by a charter. For the nascent property owner, the act provided for the acquisition of individual holdings, labelled 'individual ranches'.

Over the years the policy has been applied to virtually the whole Maasai region and parts of Samburu, so that most of the traditional communal lands in these areas have now been registered as either group or individual ranches.

But an important official practice, with a strong impact on pastoral tenure, has emerged outside of any clear legislation. This is the sub-division of group ranches into small parcels simply called 'ranches'. This started in 1982 with a dispute between owners of a group ranch in Narok district about the rights of inheritance of members with different numbers of sons. When the government resolved the conflict by supervising sub-division, a precedent had been set which has allowed sub-divisions to spread across the district. This has occurred in the absence of legislation or, indeed, of any developmental framework.

Although some forms of group ownership have been attempted in northern Kenya, the tenure changes emerging from government practice mainly predominate in the southern cluster of Kenyan pastoralists.

TRENDS IN PASTORAL DEVELOPMENT

To appreciate the dynamics of tenure change, it is useful to outline some traditional tenure relations among the pastoralists and to examine how policy has affected them.

The Boran

The Boran are a pastoral people living in southern Ethiopia and northern Kenya. They keep cattle, sheep, goats and camels. Livestock is owned and husbanded by the household unit. This comprises a man, his wife or wives and their children. Access to pasture and water resources is achieved through belonging to clusters of villages in a neighbourhood (*ardha*).

Traditionally, several *ardha* combine to control access to a pasture zone (*deda*) and regulates the seasonal husbandry of different species. This grazing zone is managed by the elder heads of the constituent villages under the nominal leadership of one elder selected from among this group.

Although access to the *deda* is not strictly limited to residents in the villages which control it, the management unit of each *deda* can exclude outsiders, and can apply its regulations to all herders using the zone.

Although pasture is an important and often scarce resource in the dry Boran territory, it is the watering points for livestock that are critical in the region. These are regulated within a larger territorial unit (*madda*). As the wells are scarce and constitute major investments in time and labour, their use is carefully controlled by a *madda* council which assigns maintenance and protection duties to those who use them.

Some eminent elders in prominent families also marshall labour and build their own wells. Under this system, the individual ownership of water points does not extend to the control of pasture in the neighbourhood. However, by controlling a critical resource like water, such families do, in fact, regulate access to adjacent pasture. This contrasts with the *madda* system where well management also manages pasture (Baxter 1978, Legesse 1973, Helland 1980).

Clearly the control of water points is pivotal to land tenure for the Boran. Access to water sources would underpin any change in land use. The rich and powerful elite in Boran society have tried to enhance their status by investing in boreholes to secure permanent private water sources. Traditional water systems relied on collective effort, but individual boreholes, sunk with hired labour, have narrowed the range of community sharing. One of the consequences is a growing inequality among the Kenyan Boran and the establishment of a link between urban and rangeland investments by the rich.

The government has addressed the water problem by sinking public boreholes. This has taken resources away from both the traditional 'big men' whose clans own wells, and from the *madda* use. The result has been a proliferation of permanent settlements near water points, and the over-grazing of the land adjacent to these water sources. The subsequent and inevitable degradation of these areas is emerging as the most serious environmental problem in the region (Lamprey and Yusuf 1981).

While tenure changes have been justified as incentives to more responsible resource management and enhanced capital, there is no clear evidence of improved animal husbandry among the Boran. No study has yet demonstrated increased revenue resulting from changed access to water and pasture.

The Turkana

In Turkana district, to the northwest of Kenya, changes in land tenure have been late and modest. Traditionally the Turkana evolved a tenure system which responded to the dry conditions of their territory and recurrent drought—the social organisation permitted extensive movement between resource areas, from season to season and from year to year, and regional links with other pastoral groups were developed.

The long distances that needed to be covered to sustain a living in such severe conditions meant that local groups did not have full control over resources. Their enduring social organisation revolved around the mobile herding and livestock protection party (*Adakar*). The focus of the clan was the herd, not local pasture resources.

An important tenurial factor among the Turkana is the practice of 'tree tenure' whereby individual families own primary rights to some fodder trees for use in the dry season.[1] This crucial resource during periods of failed rain is usually in the vicinity of the relatively permanent settlement (*ere*). Here women, children and old men remain with weak and lactating cattle (Hogg 1986).

Government intervention in Turkana has constrained the operation of traditional tree tenure arrangements. One important step was the regulation of user rights to trees as an attempt at resource conservation. The other is the lack of recognition for *ekwar* — preferential user rights traditionally held by individual families over trees with fodder value. Developments, such as the damming of the Turkwel river, have affected the tree cover around the riverine forest, further complicating the traditional usufruct arrangements.

The pattern of settlement in the famine relief camps provided by relief donors, NGOs, and government has ignored traditional tenure arrangements around trees in favour of convenience and practicality. Local pastoralists have thus been denied an important resource. The location (within Kenya, Turkana only borders the agro-pastoral Pokot and Marakwet), combined with many years of poor communication, have kept the area free of infiltration and competition from cultivators. For many years the main infiltration was from Somali livestock traders (see case study 2).

In the 1970s the construction of an asphalt road through the district and the establishment of mechanised, irrigated agriculture at Katilu in the south ended this insularity. This project, funded by the government, NORAD and the FAO, involved the damming of the Turkwel River and the gravitational irrigation of plots, which were sold to applicants.

The scheme entailed the private ownership of land with a title deed. All plot owners were contracted to cultivate cotton and horticultural produce for the down country market. Very little food was produced, in spite of Turkana's increasing dependence on maize. The traditional production (by women) of sorghum in small riverine plots continued.

Although the Katilu scheme has not succeeded, it has influenced the use of range resources among the Turkana in several ways. It was the first case of substantial alienation of pastoral land in the area, and has been followed by similar privatisation of lands, especially in the neighbourhood of the emerging trading centres. It also pre-dates the much larger Turkwel irrigation project, which is likely to convert more land into cultivated fields under private title.

The Katilu scheme also brought whole families of peasants from other parts of Kenya into the district to become plot owners. Many have settled permanently, contributing to the emerging commercial centres and experimenting with dryland cultivation.

Since the 1970s the Catholic Church has increasingly expanded into education and health support. Since 1980 the Turkana Rehabilitation Project, funded by the Dutch government and the European Union (EU), has evolved from famine relief into drought monitoring and contingency planning. It also focuses on improved livestock management. Development assistance to Turkana has generally been 'drought-driven', rather than consisting of attempts to improve the pastoral sector (Helland 1987).

New developments, such as irrigated agriculture, fisheries development along Lake Turkana, and food-for-work projects around the relatively permanent famine relief camps, have had a substantial impact on social relations in Turkana society. One important change is in the nature and composition of families. For settled families without capital who want to return to pastoralism access to cattle for bridal payment is a must. This in turn entails sustaining ties with people still in pastoral production. Such ties include exchanging agricultural produce for livestock, offering daughters in marriage, or retaining herds on the irrigated land in southern Turkana.

The Kenya Maasai

Of all the areas of Kenya's rangelands, the two Maasai districts of Narok and Kajiado have experienced the most dramatic changes in land tenure. The Group Representatives Act of 1968 and competition for a range of resources have affected them more than any other group of pastoralists.

Traditionally, the Maasai were organised into large sub-tribal groups (*olosho*) which occupied specific territory. Totems, dress and ceremonial rites differentiated each group. Most importantly, an *olosho* laid exclusive claim to territory. The territorial sections were then divided into locality units (*enkutoto*) which contained a cluster of local communities.

In the traditional herding system livestock are owned individually in the name of the head of the household, while pasture, water and salt-licks are communal resources controlled by the *enkutoto*, which could exclude outsiders. The *enkutoto* constituted the minimal ecologically viable unit for all-season pasturation.

The thinking among an elite group within the Maasai—as well as within the new government in the 1960s—was geared towards privatisation. This trend was accentuated by the Anglican Church when it established the Maasai Rural Training Centre at Isenya in 1961. This is a multi-purpose centre whose most consistent theme has been the inculcation of entrepreneurial and commercial skills among the Maasai. Many of the early agitators for the sub-division of land and titling as private property came from this centre (Kituyi 1985).

Under the new tenure policy most land in Narok and a substantial portion of Kajiado have been transferred to private ownership. While the first phase (into the late 1970s) saw a predominance of group ranches, the creation of some individual holdings for a property elite fuelled a transition to further sub-division.

The official policy since the 1960s has been to model group ranches as closely as possible on the traditional local territorial unit, the *enkutoto*. However, in practice little regard has been shown to traditional access rights within the pastoral community. Many group ranches, especially those created since the late 1970s, are merely geographical units occupying space between other adjudicated lands.

Individual ranches—early beneficiaries were relatively influential—received some of the best land available (White and Meadows 1981). Whereas group ranches could be tailored to traditional user groups, the creation of individual ranches in an area with a strong community-based culture represented a clear departure from tradition.

In areas of western Narok, private claim to traditionally communal land has been established and recognised by the community. This involves an average of 20 hectares (ha) per household in East Mau, and 10ha per household in Mulot and Kilgoris (Government of Kenya 1986a).

Rich rain-fed lands seen as 'unoccupied' have been gazetted into forests covering 270,550ha of Narok. These include 150,000ha of private forests, mainly in the water catchment areas. Major harvests for timber and charcoal took place in the late 1980s.

Many Maasai pastoralists contend that the immigrants pose less of a problem than locals who have been allocated prime portions of the rangelands as individual holdings. Many state that they have given into the ranching policy because their leaders were 'taking all the land'. Among the Samburu, many group ranch committees have reverted to short-term leasing of their lands to urban wheat farmers, as has happened in Narok. Some arrangements have been extended for so many years that the lands have virtually been lost to the pastoral economy.

Among the pastoralists of the southern rangelands, individual ranches have restricted herding arrangements. Group ranches tend to allow for the interranch movement of livestock, broadly in the vein of traditional resource use, but the boundaries of individual ranches are more rigidly enforced. This tends to force the poor into more extensive herding in order to compensate for the loss of prime dry season pastures to the neighbouring individual ranches.

In Narok and northern Kajiado, sub-divided group and individual ranches have been extensively sold to non-Maasai peoples. Narok has been a major destination for immigrants from central and western Kenya, so much so that now these immigrants account for over 50 per cent of the district's population. This influx is at the core of a conflict which has garnered nationwide political attention (see case study 1).

Large companies have also invested in farming in Narok. By 1979, East African Industries (Unilever) was growing 700ha of rape seed in the district. This area increased substantially in the 1980s. Kenya Breweries were growing 16,500ha of barley in 1980, and this too has greatly increased since then. Wheat cultivation has been undertaken by farmers on an ever increasing area of leased land. Today Narok is the main wheat-growing district in Kenya, and the main source of barley for the brewing industry (Government of Kenya 1980—1988).

MAIN AREAS OF CONFLICT AND ALIENATION

Traditional Regulatory Mechanisms

There used to be important traditional mechanisms for regulating individual conduct among the Maasai. Local committees controlled access to resources and restrained aberrant behaviour, and the eminence of traditional ritual leaders, particularly in the resolution of conflicts, were part of this traditional regulation.

The new ranch system has replaced these forms of authority. The power to resolve a conflict over the use of resources has now passed to a group whose mandate does not derive from traditional respect for the elders, but from members' elections and external patronage. Whereas elders' councils had a broad social mandate, their replacement is not competent to address wider social issues. This is manifest in the increasing recourse to official litigation in cases which could traditionally have been resolved locally.

One of the most important traditional institutions among the Maasai is that of 'stock associateship'. This mechanism of cattle loaning ensured the temporary transfer of animals from households with relative abundance (and low labour resources) to those with few cattle. It especially helped victims of drought and epidemics. The creation of ranches, often with rigid boundaries, hinders the flow of such resources. Another negative trend is the reluctance by groups, which have invested in improving resources, to allow one of their own members to borrow another person's cattle from outside the ranch. This is another way in which the tradition of associateship has been superseded.

Survival Mechanisms

The ranch system ensures unequal access to critical assets like water points, salt-licks and dry season pasture. Those who own individual ranches have grown richer and minimised their vulnerability to drought and seasonal variations in resources. These people depend on the market for storing surpluses beyond their capacity to manage, and for hiring additional labour, if required.

Control of range resources offers an opportunity to turn seasonal windfalls into capital by investing in fodder, insecticides and water. Seasonal variations in returns from investment are institutionalised by conversion into relatively permanent resources. This reduces the need for traditional solidarity, and leads them away from the traditional accumulation of goods as a way of acquiring status and prestige.

The emerging disparity in wealth among the Maasai affects their capacity to survive and recover from crises. A study from two group ranches in southern Kajiado (Grandin, De Leeuw and Lembuya 1989) has shown that richer pastoralists survive better in an extended drought than do their poorer neighbours. The rich are able to prepare their animals for the dry spell by the early introduction of a two-day trek to water sources, alternated with pasturation. This reduces the milk supply, something which the poor can not afford to attempt. The cattle of the rich are therefore well prepared for long, waterless treks when the drought becomes severe. When the poor eventually have to move they lose many cattle that have not been acclimatised to living without water.

Rich herders, especially those with individual ranches, can also secure relatively rich 'fallback sites' where they send their cattle in times of drought. This combines financial incentives and the availability of hired labour to split and herd the cattle. Both of these factors work against the poor, who depend on free access to park fringes during droughts and who cannot afford the additional labour.

Social Equity

Several emerging trends in the ranch system work against traditional egalitarianism. Disputes over intergenerational transfers of rights among group ranch members with different family sizes cast doubt on the assumption that new tenure brings security. Poorer members have been known to lose part of their share to descendants of rich group ranch members. There are reports that people who have left the area to seek work have returned only to find they have been disinherited. Some committees have denied inheritance to the sons of members when they have been away from the ranch for long periods.

In the case of individual holdings, there have been real differences in the methods of inheritance. Some men sub-divided their land equally among all their unmarried sons and daughters, others among their sons only, while others left land to the oldest son only. Each of these methods of transfer could be supported by traditional precedents, but in the case of the last two methods inequity through age and gender have become a reality within the household (Kipuri 1989).

Women are the worst affected group in the transition from traditional to individual tenure among the pastoralists. Among the Maasai and Samburu, women's rights were not considered when the ranch system was introduced. In the early stages of group ranch adjudication, all adult men and widows were registered as members. This included those who no longer resided in the area, such as migrants to towns. Registration of members was according to family. This way, women and young men, while not registered, were assumed to be part owners in the same way they had been benefiting from family resources before changes in tenure. However, there was no specification for unmarried, separated and divorced women. Exclusion from membership was quite arbitrary and was determined by the ranch committee on the basis of the age set one belonged to.

As the shortage of land became more acute in the decade that followed adjudication, unequal access became more prevalent. In some cases younger men were pushed to more marginal areas of the range, as the elders assigned themselves the better-watered part.

The new tenure system among the Maasai has shifted the emphasis away from social esteem and the accumulation of livestock as the means of acquiring wealth and power, towards fixed assets. Traditionally, prestige was gained through social involvement. The new system depends on exclusivity and relinquishing this 'costly' social exchange. The precious respect won by being good with people and animals is being replaced by forms of material wealth which can simply be inherited.

These trends are emerging clearly among the southern pastoralists, where tenure changes have been strongly entrenched, but little is known about the northern pastoralists in this regard. Research might help to identify ways of reducing the social cost of such change.

ENVIRONMENTAL IMPLICATIONS

One of the main environmental impacts of changes in land tenure can be traced to the encroachment of commercial cereal cultivation into the rangelands. Their conversion from protein-producing pastoral lands to cereal growing sustains more people, but at an ecological price: the replacement of a complex community of grasses and trees for a monoculture of one annual grass. Cereal cultivation also tends to extend into more marginal areas, whereas pastoralism will not use up soils which do not sustain pasture (Lane 1990, 1996a, Loiske 1990).

A pastoral economy is primarily the pursuit of food security through the management of livestock, and only secondarily the regulation of pasture use. Consequently, when pasture has been alienated or human concentration requires high stock numbers, attention may be diverted away from range management to the herds, so contributing to ecological degradation (Hjort af Ornäs 1989).

It is at least partially true that overgrazing—the depletion of future pasture—is the most obvious form of resource degradation among Kenyan pastoralists today (Livingstone 1977). How to measure it and how it can be prevented are not so clear. But degradation in many areas of Kenya's rangelands may have more to do with changed conditions than with traditional practice.

CASE STUDY 1: 'IMMIGRANTS' AND THE MAASAI

Some senior politicians come from a pastoral background. The land question has become an important political factor in the Maasai districts, as it becomes more scarce, and the numbers of immigrants grow. As a result, leading politicians have been under pressure to stand for 'Maasai rights'.

In May 1991, the vice-president (a Maasai) successfully campaigned for the nullification of the title deeds for 20,000ha of Maasai land which had fraudulently been allocated to senior officials in the Ministry of Lands and a few leaders of group ranches. Earlier, in February, the then minister for local government, Mr William ole Ntimama (who is also a Maasai), caused a major controversy when he declared at a public rally that all agricultural immigrants in the district should 'lie low like an envelope' or else face the wrath of the Maasai.

This was not the first time the powerful minister had threatened bluntly the immigrant peoples in Narok District. In 1989 he announced at another public rally that the Maasai, not having the advantage of education, should rightly arm themselves with spears (which he equated with pens) to drive immigrants from their lands. This was followed by a spate of night raids on immigrants' homes.

The minister's remarks have been very popular with the Narok Maasai, who have seen the numbers of immigrants swell beyond the Maasai population, and take over some of the best land in the district. Indeed, the minister has been strongly supported by other politicians from pastoral communities, with the most eminent Turkana politician asking all Kenyans to go back to their ancestral homes.

CASE STUDY 2: SECURITY AND RESOURCE TENURE AMONG THE TURKANA

The Turkana have had mixed relations with their nomadic Karamoja, Jie and Toposa neighbours in surrounding countries. This used to be regulated by the need for long distance pasturage (Map 5), but cattle raiding has always been a reality.

Over the past two decades, wars in Uganda, Sudan and Ethiopia have ravaged the economies of the nomadic neighbours of the Turkana, while also providing access to very sophisticated weaponry. This has led to an increase in raids and the devastating use of firepower. The frequent raids have made the Turkana virtually abandon the northern half of their district, although it is relatively wetter than the central region where most people now live.

Government attempts to open water sources in the northern region to utilise the abundant pasture there have been resisted by the Turkana. They argue that the main weapon for preventing the Toposa of southern Sudan from stepping up their cattle raids is the absence of water over a wide buffer area. They contend that raiders cannot march cattle for the four days it takes to get from central Turkana to the Sudan border, without watering the animals. Opening new water sources would remove this barrier and facilitate more devastating raids from the north.

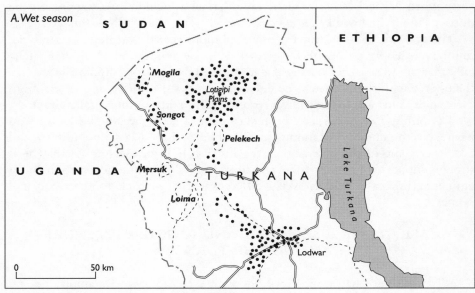

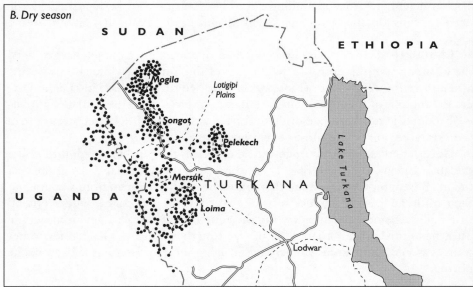

Map 5 Dispersal of Population in Northwest Turkana

CONFLICT RESOLUTION

In Kenya, most conflicts are resolved in the official courts of law. Although local authorities are supposed to be able to resolve claims, any party can bypass them and go to the courts which wield higher authority. This disadvantages the poor and illiterate, because a lack of awareness of court procedures often mean failure to show up on time, if at all, which leads to the complaints being dismissed.

In many pastoral communities, the age group system still operates and, together with the traditional elders' council, resolves conflicts over resources at the local level. Members of elders' councils interviewed by the author have admitted to increased numbers of conflicts over land, especially complaints by traditional rightholders who have been displaced, and cases of trespass (Kituyi 1990).

Among the Maasai, especially of Narok, the new tenure practices, which have permitted massive immigration, have provoked different conflicts and forms of protest. Local pastoralists have deliberately released their animals onto cereal fields, and more recently there have been armed incursions by groups of Maasai youth. These protests relate to a sense of injustice for a development that is perceived as a threat to individual security. As Raintree asserts: 'if a [development] plan unwittingly increases the ambiguity of my relationships with others over that parcel of earth I depend on, the chances of my cooperation decreases in direct proportion to the insecurity created' (Raintree 1987).

CONCLUSIONS AND RECOMMENDATIONS

The changes in tenure and trends towards greater commercialisation have caused, or have been accompanied by, a marginalisation of most pastoral societies in Kenya. This works against sustainable development, which can be described as 'seeking to meet the needs and aspirations of the present without compromising the ability to meet those of the future' (World Commission on Environment and Development 1987). The declining economic security of pastoralists makes them vulnerable to the vicious cycle whereby 'poverty lead[s] to environmental degradation which in turn leads to even greater poverty' (ibid).

While preliminary evidence points to resource degradation following changes in tenure, official thinking remains grounded in the 'tragedy of the commons' argument that common land is inevitably degraded by overuse. The issue for researchers should be the circumstances under which the equilibrium between personal gain and social regulation is disturbed. What makes some pastoralists in a common property regime behave as if operating out of an open-access regime? One possible explanation for the decline in social regulation and the increase in competition is an increase in human and livestock populations. It is argued that in common property regimes with finite boundaries and limited human and stock offtake, an increase in the number of 'rightholders' will deplete more resources than natural regenerative capacity will be able to replenish.

For a long time the relationship between researchers and service delivery groups among the pastoralists has been very weak. Only recently has this problem begun to be addressed by groups such as Nomadic Pastoralists in Africa (NOPA), operating out of the United Nations Children's Fund (UNICEF), and the Oxfam pastoral network, both based in Nairobi. The new Pastoral Tenure Network, which was formed by researchers and NGOs from eight East Africa countries, operates out of the African Centre for Technology Studies, also based in Nairobi. This needs to be strengthened. The following are the most important issues that need to be considered and addressed:

1. Despite growing interest in the balance between people and their environment in pastoral areas, the range of research in this area remains modest. One clear gap is the dearth of knowledge on what is happening among the northern pastoralists.
2. A re-examination of the justification for land registration is now needed in the light of increasing evidence that changes inspired by the government and donors have not yielded the desired results. The claims that land registration increases productivity warrant urgent attention.
3. If the claims used to justify land registration are invalid under a broad range of circumstances, then current trends in official practice should be judged either premature or inappropriate, even without considering registration's contribution to social inequality (Atwood 1990).
4. A great deal of indigenous pastoral knowledge on sustainable resource husbandry has been ignored. Consequently, policy research must begin by redressing this. Understanding how people managed resources in the past will help in gauging the limits of traditional solutions to today's problems of land use and the environment.
5. Specific areas of research should include:

- the relationship between the pastoralists' need for secure long term development and the short-term nature of most development interventions;
- the changing balance of power;
- the impact of unequal access to critical resources.

6. The research agenda also needs to examine how to strengthen the capacity of pastoralists themselves to seek justice. In pursuit of this, donors could award

scholarships to individual pastoralists, so that they could contribute actively to solving their people's problems.

---

[1] My thanks to Edmund Barrow of AWF for helpful information on this topic.

3

MALI

Richard Moorehead
Research Associate, Drylands Programme, IIED

Map 6 The Republic of Mali

INTRODUCTION

This chapter focuses on the inland Niger delta which contains some of the most valuable pastoral resources in the Sahel, and which provides sustenance to nearly one million cattle and over two million sheep and goats during the dry season each year. This equates to 20 to 25 per cent of the national herd.

The delta, located in central Mali (Maps 6 and 7) is an area of international environmental importance in the Sahel. It contains fisheries, pastures, agricultural land and forests, all of which are under significant threat from 20 years of lower than average rainfall and flood levels, and probably from increasing population pressure and intensifying human use. In recent years, dry conditions have forced rural producers to increase exploitation of fisheries, pastures, forests and wild food resources as production systems have had their activities diversified in response to drought. Outsiders to the delta have moved into this relatively rich area in the northern Sahel, which acts as a safety net for producers in surrounding marginal millet-producing zones, drylands pastures and forests.

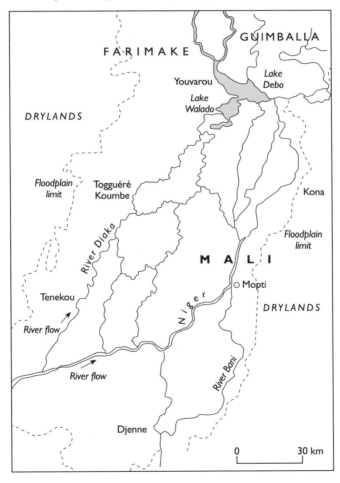

Map 7 The Inland Delta of the River Niger

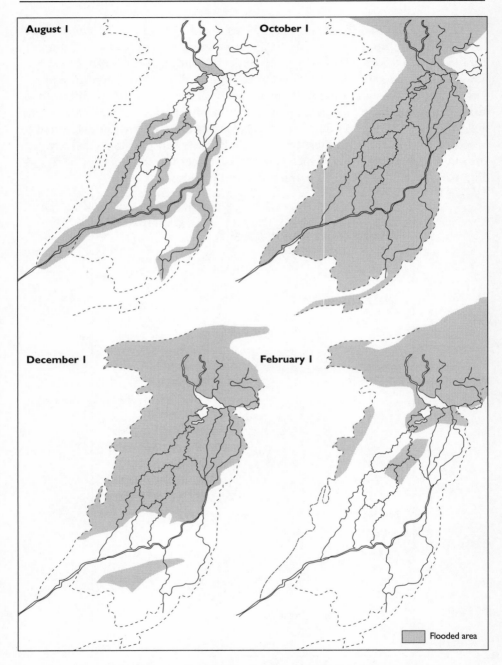

Map 8 The Seasonal Flood Regime of the Inland Niger Delta

The diversity and productivity of ecosystems found in the delta owe their existence to the annual flooding of the area, which transforms the zone from a dry plain into an inland sea at different seasons of the year (Map 8). The inundated plains can cover anything up to 25—30,000km^2 of land, depending on the amount of floodwaters descending the Niger and Bani rivers, which have their catchments 1000km to the west and southwest of the delta where rainfall is between 1300—1800mm. The delta spans the Sahelian zone to the north, with 200—300mm of rainfall, and the Sudano-Sahelian climatic zone to the south, with 500—600mm a year. The rainfall pattern and the timing of the floodwaters provide the seasonal cycle of the area: between April and June the area is hot and dry, with the water running only in the deepest water courses; between June and July and September and October the rains fall and the water rises to spill over the floodplains; between October and January the area becomes an inland sea; and between January and March the water falls in the delta until it reverts to a dusty plain.

In addition to considerable wildlife resources, including 130 genera of fish, 300 species of birds, and the presence of large mammals such as manatees and hippopotami, the delta contains rich agricultural land and some of the best dry season pasture found in the Sahel, well-known as the *bourgoutières*, which are floodplain pastures capable of growing in three metres of water and of producing up to 25,000kg/dm/hectares/year (kilometres of dry matter per hectare per year) in optimal conditions (IUCN, 1987).

These pastures form the hub around which the pastoralists of the zone design their pastoral cycle: they leave the delta for the surrounding drylands at the onset of the rains, sometimes moving 200—300km in a loop before moving back onto the floodplains in October and November. Between then and January and February the herds move northeast down the delta following the flood pasture as it becomes available as water levels fall. The regions of the lakes around Debo and Walado are the dry season refuge for the herders before they renew their cycle with the arrival of the rains.

The larger part of these herders are Fulbé (Fulani) numbering perhaps 100,000 people. From the thirteenth century onwards they arrived in the delta in successive waves from the west in nomadic groups called *ouro*, led by their shepherd leaders, the *ardo*. There are four main groupings of Fulbé in the delta: the Dialloubé, the Ouroubé, the Fittobé, and the Ferobé, of which the most important are the Dialloubé. These Fulbé groupings moved into the western and northwestern reaches of the delta first, before spreading over the northern edge to settle on its eastern border (Gallais 1967).

As the Fulbé moved into the floodplains the clans fragmented, and small groupings from the four major clans either carved out smaller territories or became mixed groups with access to defined pastures. The human and social geography of the delta Fulbé still retains the characteristics of this process: the four largest and most homogeneous pasturing territories remain in the north and west, and the smallest, most fractured ones are found in the east and south.

Apart from their differentiation by clan, the delta Fulbé are of two types: so-called 'red' and 'black' Fulbé. The former are the descendants of the original nomadic groups, specialised in the transhumant management of animals. The 'black' Fulbé are descendants of marriages between *ardo* and cultivators who are the oldest inhabitants of the zone, called the Marka. The offspring of these alliances created a class of Fulbé with interests both in pastoralism and the cultivation of millet and rice. They follow a more sedentary lifestyle in villages, and only join their animals when they return to the floodplains after transhumance. Nowadays they are typically the heads of Fulbé villages, and safeguard the floodplain pastures while the animals are away from the delta.

As with other Sahelian pastoral groups, early on the Fulbé developed master—servant relationships with agriculturists in the delta in order to provide them with a reliable source of grain. This 'slave' ethnic group, known as the Rimaïbé, is composed of people belonging to several local clans—Bambara, Bobo, Bozo, Dogon and so on—and using the floodplains and neighbouring drylands in sedentary cultivating communities.

POLITICAL AND LEGAL CONTEXT

Between 1820 and 1862 the entire delta was unified into a theocratic Muslim state under the hegemony of the Fulbé, who managed the area in line with their predominantly pastoral interests. Much influenced by the kingdom of Sokoto (Barth 1890), Cheikhou Amadou, spiritual leader of a federation of Fulbé clans in the delta, obliged the inhabitants of the zone to fix an area of origin, build permanent dwellings, and take up the Muslim faith. Pastoral, farming and fishing territories were codified and 'slave' cultivating settlements composed of Rimaïbé were settled on the floodplains and the surrounding drylands. This period is known in the delta as the *Dina*.

Through 70 years of French colonial rule and 30 years of independence as part of the Republic of Mali, the *Dina* system has retained its importance, not only for the allocation of pastures, but as the reference point for allocating access to agricultural land, forests and fisheries in the delta as well. It is therefore essential to have an adequate understanding of the system in order to appreciate the profound changes that are now taking place in the area and the pastoral tenure system.

The evolution of property rights to pasture obviously preceded the *Dina* period and was based on the continual movement of Fulbé herds through the area. Access to pastures in this earlier period, known as the period of the *Ardubé*, was through the clan head, the *ardo*. The *ardo* claimed direct male descendance to the founder of one of the four Fulbé clans of the delta. The right of the Fulbé to the floodplains was seasonal: when the floodplains were covered in water, rights to exploit the zone belonged to the fishermen and farmers of the zone. During the dry season the *ardo* decided the dates of entry of herds into their pastures, received rents from, and authorised strangers' access to, the zone, and fixed the dates of exit. Access to pasture was therefore subject to kin links or to payment, and the *ardo* was the sole authority empowered to grant entry. Oral accounts seem to indicate that pasturing territories during this time were largely independent. There were reciprocal rights to pasture, but each territory was managed as an autonomous unit and permission from the *ardo* was sought each year for entry into their pasture.

As waves of migrants continued to arrive and the Fulbé began to intermarry with local people, a more detailed management system for pastures evolved. Settlements of Fulbé were established in which the 'black' Fulbé lived—known as *ouronké*—within more closely defined pasturing territories. At the same time their *ardo* established farming communities of Rimaïbé in the delta. Between the fourteenth and seventeenth centuries power devolved to two different groups within each clan: the *ardo*, with overall power, but particular responsibility for the pastoral interests of the clan, and the head of the Fulbé village, or the *ouronké*, known as the *diom-ouro* or *dioro* for short, with responsibility over the cultivating communities and for village pasture land where milk animals were kept when the other animals were on transhumance, known as *harrima* (Gallais 1967).

The *Dina* system of the nineteenth century basically codified this system, but at the same time suborned the independence of each territory to the collective interest of the delta-wide state, and in so doing defined formal linkages between Fulbé clans and between other production systems using the area.

The major transformation brought about by the *Dina* was the forced sedentarisation of the delta's inhabitants, and the definition of the resources to which they had access. Pastoral resources in the delta were divided into approximately 37 territories, known as *leyde* (singular *leydi*). These territories were, for the most part, allocated to their traditional clan managers, but with important differences as the power of the *ardo* was broken by military force at the outset of the *Dina* by the federation of clans supporting the Islamic state of Cheikhou Amadou. The *ardo* of the Macina Kingdom on the western border of the delta were bereft of much of their power and settled at Ténénkou, on the Diaka river. In areas of the delta where Fulbé clans were animist (the Dialloubé, for instance), pastures were given to religious authorities or *marabouts* on the frontiers of the territories of these groups as surveillance posts. In pasturing territories formerly managed by *ardo*, *marabouts* were often installed as political leaders, leaving *ardo* lineages the responsibility for managing pastoral activities.

The *Dina* system codified the transhumant pastoral system in far greater detail than before. Routes into and out of the delta were fixed; dry season itineraries on the flood pastures and camping sites were defined; and the membership of herd associations within clans for the dry season transhumant cycle were established. At the same time the dues payable by the Rimaïbé to their masters were fixed, as were levies on independent farming and fishing communities. Rules governing the management of herding over the whole delta were created: the order in which herds crossed into flood land pasture, and the places in which these crossings were to take place, were fixed; pasture in the whole area was decreed open-access when the *Bal-mal* constellation of stars (the Pleïades) in the Islamic calendar appeared each year (mid-March).

Under the *Dina* system herds belonging to the same territory, or *leydi*, were divided into transhumant groupings known as *eggirdé* (singular *eggirgol*), with each *eggirgol* (10 to 60 herds approximately in each one) led by the head of a sub-lineage of the founding clan, the herds belonging to its constituent members. *Eggirdé* were created essentially as protective herding units against the depredations of the Bambara and Kel Tamasheq found in the wet season pastures neighbouring the delta, and formed up to leave when the rains started each year. Clan territories had reciprocal access agreements with other territories in the delta, thus allowing clans from downstream pasturing areas access to upstream and central territories, and vice versa. Each territory had a number of entry points (most often crossing points on a river, known as *deggê*).

On their return from transhumance at the end of the wet season, *eggirdé* would form up in the waiting zones, according to their access agreements, beside the two great crossing points leading to the first pastures upstream to be revealed by the shrinking floodwaters. When the *dioro* in the first territories gave the order, they moved into the pasture in strict sequence, with the owners of the pasture first, followed by the animals of the other territories with whom they had reciprocal agreements, followed by the animals of strangers to the delta who had to wait for some days before they were allowed in. The *eggirdé* of each clan, and herds within the *eggirdé*, moved into the pastures in a specified order. At the end of a series of *deggé*, as the waters fell, the herds moved on to the lake beds in the north of the zone. Some three weeks after the last crossing, pastures all over the delta became open-access.

The evolution of the *Dina* system can be seen as a shift of power away from permanently mobile groups (represented by the *ardo*) towards sedentary communities (represented by the *diom-ouro* and the *marabouts*) within a hierarchy that made the pastoral economy predominant over fishing and farming interests. Every square metre of the delta and its neighbouring lands had an owner and a manager, with the pastoral manager of a territory as the overall head of all the communities living within his zone. Under the *Dina* three forms of collective property existed. *Beit-el*, basically state land, was either administered by *marabouts* or run directly in the interests of the state from Hamdallahi (the capital of the *Dina* state), and which included property of those who died intestate as well as resources allocated for the payment of administrative costs. Village pasture provided feed to milk herds left behind when the other animals were on transhumance (*harrima*). Cultivating land and collective pasture of the clan were reserved for transhumant herds' use. These last three were managed by the *diom-ouro*.

The *Dina* administered the area in favour of the collectivity of all the *leyde*, rather than for any particular territory. This was coordinated through the maintenance of a standing army to protect the state's boundaries, paid for by fees and taxes levied on each territory, and through the organisation of defensive groupings of Fulbé herds while the animals were on transhumance in the drylands. Furthermore, the codification of fishing, farming and herding territories and the establishment of 'state' land provided a comprehensive set of rules of access to resources in the delta, while defining the boundaries of each terrritory more closely than ever before.

TRENDS IN PASTORAL DEVELOPMENT AND WELFARE

Resource Management and Tenure

The French generally accepted the boundaries that had existed under the *Dina* system, and administered the delta through a set of '*cercles*', divided into 'subdivisions', and at the lowest level '*cantons*' (about 35 in all). While the *cercles* and sub-divisions included territory the *Dina* system had not governed before, the *cantonments* in the delta broadly reflected the centres of local power in the nineteenth century. From the point of view of local producers the hierarchy of power remained for the most part the same up to the sub-division level—that is, local land and fishery resource managers liaised with the *chef de cantonment*, who looked in turn to the head of the sub-division and the *cercle*. Generally, the heads of *cercles* and sub-divisions were French expatriates; those of the cantonments were the traditional chiefs whose families had been originally appointed under the *Dina*.

While the French accepted in some measure to rule through indigenous structures, the centralised power of the *Dina* system, based at Hamdallahi, was dismantled. Of great symbolic and some economic significance was the formal abolition of the status of slaves negotiated between the French and the *dioro* and *ardo* for the Rimaïbé in 1903. Reports from the time speak of thousands of Rimaïbé leaving their original settlements and moving to other communities after the treaty was signed (Gallais 1967).

Though customary managers were left in place by the colonial authorities, their ability to enforce access rights and to manage the full range of resources they traditionally presided over was significantly undermined by the French, whose policy with regard to land rights was to lead directly to the insecurity of tenure that characterises the region today. Like many other colonial regimes in Africa, the French regarded land left under longterm fallow and other resources only used seasonally as underutilised, and that this could be more productively developed. In line with the practice in their own country, the French believed the best means of achieving this was to allocate what they termed 'vacant lands' to individuals as private property or to manage it as state property.

In order to do this, the French state had to own these resources. Thus, in 1904, the French colonial administration brought in legislation that decreed that all vacant lands were henceforth the property of the state (*domaine privé de l'état*). Concerns or individuals could gain title to this state land if they drew up a detailed development plan of:

- how the land was to be developed;
- how many people would be employed on it;
- who local people living on the concession were;
- the areas to be set aside for their crops, etc.

When the colonial administration considered that the work to be carried out under the development plan was completed, the concession became the property of the owner (République du Mali 1987a).

The effect of this legislation was to set up two parallel forms of land tenure: the one customary, governing areas under continuous production (as defined by the French), and managed by customary authorities; the other written, conceived as being 'progressive', covering vacant lands to which private title could be gained and sanctioned by colonial legislation. This distinction was informed by a Eurocentric notion that 'vacant' land encompassed land that was not cultivated (including fallow land that had been left to regenerate for longer than ten years), and in particular included forests and pastures.

The policies and laws of the colonial state affected substantially customary pastoral tenure systems and management regimes of the delta. Of seminal importance for the future was the fact that the final arbiter of disputes was now an entity that was foreign to the delta, and to the region as a whole (the French

Colonial Service). The primary objective was to extract revenue (taxes), goods (rice, wool, hides and so on) and services (military personnel) for the colonial state. Whereas under the *Dina* system the area had been administered by a political and economic structure based on the delta, and rooted in its history and customs, the French administered the zone as one small (and fairly unimportant) part of Afrique Occidentale Française. Of real practical importance was the separation of resources into those owned and managed by the state (pasture, forests, wildlife, fisheries), and those remaining in local people's hands (agricultural land). This led to some local communities, and groups within settlements, having a vested interest in promoting state ownership over customary rules. This afforded them new access to the more productive resources denied them under the former system, but which were available to them if they were zoned as state land (Moorehead 1991).

The colonial state actively intervened in the management of resources under customary control, particularly in the case of pastures. Colonial administrators regarded the downstream grazing lands around the lakes as a strategic reserve for livestock when pasture in other areas had dried out or been eaten. And they considered that premature entry into these pastures would jeopardise the ability of livestock to survive at the end of the dry season, particularly in drought years. During the colonial period the first elements of a system were put in place which were subsequently to be refined by the independent administration

Under this system the administration coordinated and fixed the dates at which animals entered the different pasturing territories of the delta, on a broad axis southwest to northeast as the waters fell, in conjunction with the local *dioro* and the *chef de cantonment*. The object of the intervention was to stagger the crossings in such a way that livestock could be retained in the central area of the delta until well into the falling-water season (March—April). Thus, when they arrived on the floodplains around the lakes, there would be sufficient pasture to take them through to the onset of the rains when they could leave on transhumance once more. This was effected by setting the dates of entry of some 30 principal crossings that traditionally controlled access to different territories, and policing them.

The French attempt to manage these crossing dates both paralleled and changed the customary system. According to herding managers in the delta, its principal effect was to undermine the autonomy of each territory: under the *Dina* system each *leydi* had decided individually as to when entry to their pasture was to be effected in line with pasture conditions each year. They informed their neighbours and the other clans with whom they had reciprocal access agreements, but they were not bound to link their dates with those of the others. Colonial policy attempted to unify this system, so that the delays between crossings were fixed once the initial date of entry at the upstream end of the delta had been decided.

This broke the chain of command that had existed under the *Dina* by which access to resources depended on the relationship of a household to the founding line of a community. In this situation the capacity of a customary manager to impose a collective decision over all other users and to enforce entry and exit rules, especially with regard to strangers, was weakened by the presence of a substantial 'veto position' held by the colonial administrator who could call on considerable force in extreme cases to back his decision (Oakerson 1986). Formerly subservient or stranger groups could appeal directly to this wider structure which, from the sub-division level up, was staffed by people who were new to the area and whose knowledge of local customs and history was perforce minimal, because they were expatriates.

Since 1960 the independent governments have built on, and extended, these colonial foundations. Since 1959, but most clearly enunciated in the present land tenure code, the state has declared ownership over all land, whether or not it is subject to customary title, is in constant use, or has an owner. This is most explicitly formulated in Article 127, under Law No 86–91 of 1 August 1986 (République du Mali 1987a).

Either individual or collective customary rights can be converted into written, or formal, rights (*concession rurale*) where buildings or other works (*mise en valeur*) have been carried out by customary holders, but, crucially, only where the resource has been initially registered in the name of the state and customary rights have been explicitly abandoned. Traditional managers can lay claim only to what they personally exploit and must not use their position to claim greater rights to land (Article 129). Thus customary rights disappear in the process of registration and traditional owners acquire simple use rights (Article 132).

The actual process of registration has been aptly described as '...organisé avec un grand luxe de détails' (République du Mali et Confédération Suisse 1987). Any registration of land over 10ha has to pass through the Council of Ministers at a national level (Article 44), all registrations have to be published in the official gazette (Article 46), registration is initially only temporary, a development plan for the land in question has to be drawn up (Article 48), and an annual fee has to be paid (Article 49). Only if all these conditions are met, and the works to be carried out in the development plan accomplished (evaluated by a commission appointed by the Council of Ministers) is registration into definitive title allowed (Articles 62/65). The state can reclaim the land at any time, although it is obliged to pay compensation once final title has been granted.

In practice neither the colonial nor post-colonial state has made any serious effort to uphold those articles of land tenure law that in theory allowed customary users of resources to register their title, either by upholding customary users' right to register land that outsiders wish to claim (Article 130), or by encouraging them to register in the first place. The Malian state, if anything, has further discouraged rural people from beginning the process of claiming title by making the fiscal payments for *concessions rurales* too high, and requiring development plans that are beyond rural people's means. Even the basic premise that local people have a right to the resources they customarily use has been weakened by the summary transfer of ownership of all resources into the state domain. In cases where a clear title to resources is of pressing importance to local people—where, for instance there is repeated conflict over access—the judicial system, responsible for adjudicating land tenure questions, rarely intervenes both because of local people's ignorance of their rights and the distance of courts from rural areas, as well as the infrequency of their meetings (République du Mali et Confédération Suisse 1987).

Of cardinal importance for herders is that their customary use of resources, which leaves little trace of their use of the land (since they move on and allow the grass and forage to grow again), is not considered to be a productive form of investment, and therefore does not allow them to claim the resources as their own, even where they would be prepared to make over the land formally to the state as part of the first condition to gain ownership. It goes without saying that almost no land registration has taken place in rural areas in Mali, and least of all in pastoral areas.

This legislation actively promotes insecurity of tenure for rural inhabitants. While continuing to exploit resources customarily belonging to the community, they perceive that the state has both the will and the power to override customary rules of allocation and exclusion. Because most of them do not understand French (the official language), and are not aware of their rights, local people are constantly disadvantaged. Agents of the administration use this legislation to impose a centrally planned management system that ignores the basic premise of the traditional regime: right of access according to kinship and residency rules. This insecurity has been linked directly to rural producers' unwillingness to participate in longterm strategies aimed at sustainable production, in favour of shortterm exploitation that yields immediate results (République du Mali et Confédération Suisse 1987). In effect, this legislation provides the foundation for the penetration of all the state's component institutions—the political party(ies), administration and technical services—into local management rules for access to and allocation of natural resources (Moorehead 1991).

In the period since independence there has been a concentration and then a dilution of centralised economic and political power over local management systems in three phases:

1. 1960—68, when a reforming Malian administration attempted to impose a 'socialist revolution' from above;
2. 1968—1991, during which a military dictatorship both widened and deepened the ability of the post-colonial state to extract revenue from the rural sector and redistribute access to natural resources between rural producers;
3. 1991 to the present day, during which time an uncertain democratic government has attempted to end a rebellion by the pastoralists in the north and east of the country and provide more popular and representative governance.

During this time the ambiguity and uncertainty surrounding tenure rights has been aggravated by two principal factors: the inconsistency with which resource management policy is implemented by the administration, and competition between the relevant political and administrative institutions of the state in order to capture considerable formal and informal revenue. Most significantly, there has been a proliferation of institutions that are linked to rural households through three major structures: the administrative hierarchy, the political party(ies), and the technical services. Cutting across these three primary structures are a set of councils and committees dealing with technical and developmental issues.

The political party(ies), the development committees and councils, the administration and the technical services of the Ministry of Natural Resources are all involved in allocating access to resources and, to the extent that they often act unilaterally and independently, allocating access on contradictory grounds. At the level of the community, this throws open the allocation of access to resources and the implementation of rules to the jurisdiction of several authorities. By way of example, seasonal visitors to a community or pasturing territory can now ask for access rights from the political party, the traditional manager, the technical services, or from the local administrator, and can play off competing interests against each other.

In the most prosaic form this concerns money, either for taxes, other charges, or for informal contributions. And this is in conditions where rural producers need access to resources in order to raise cash to buy necessities and repay debt, where officials need cash to make up shortfalls in their salaries, and where the state needs revenue to pay salaries and finance state expenditure, but where drought is narrowing the natural resource base on which they all depend.

Fiscal and Development Policy

Over the past 25 years fiscal and development policy has broken down the customary management system of pastoral resources. At the same time it has increased the revenue of the post-colonial state, and this over a period when the natural productivity of the inland Niger delta has been in decline. Since independence the Malian state has sought to take an active hand in resource allocation and management in three main ways through:

1. direct reallocation of resources between rural producers;
2. seeking to change the manner in which resources are exploited;
3. changing the way in which the value of rural production is distributed.

Customary herding territories have been transformed through the formation of new administrative areas: as late as 1977 new boundaries between *cercles* were being marked out in the delta, breaking up former *arrondissements* (the lowest administrative unit in Mali) and cantons that, as mentioned earlier, in their broad outline reflected the pasturing territories of the *Dina* state. In these new administrative areas informal authority for the management of resources has been taken over by offshoots of the original customary managers or by former dependent groups claiming access rights because they now live in a different circumscription to the original owners. This has provoked endless conflict over exactly where the division of the *cercle* lies, as the *cercle* boundaries are only partially surveyed and this is done only where and when conflict arises.

In other areas the state has directly removed natural resources from local producers' hands, as in the case of the classified forests and national parks, or has removed them from customary control and reallocated them according to external criteria. This is particularly true of the *Opérations de Développement Rurale* (ODR)—the parastatal development agencies. For example, 27,000ha of prime pastures and fisheries in the central and western parts of the delta were converted to rice polders which, following the onset of dry years, have not been very productive.

Development policy in the delta up until the early 1990s rested on the imposition of 'progressive' development initiatives on rural producers. Both the political party(ies) and the administration promote development plans that are based on sedentary activities, use high inputs of technology, and involve collective methods of production. To all intents and purposes, these ignore the special conditions which pertain to the delta, or the livelihood strategies of herders. Their plans reflect the national development policy which comprises four main initiatives:

1. the fight against desertification, including creating a 'green barrier' against the advance of the desert;
2. the promotion of village reafforestation and training programmes for local people;

3. national food self-sufficiency through the management of water resources, irrigation schemes and promotion of counter-season crops (that is, crops grown in the dry season);
4. the devolution of greater responsibility to local people for the promotion and implementation of rural development policy (République du Mali 1987b).

To these development initiatives and approaches the Malian state has added a panoply of measures aimed at controlling rural producers' hunting, fishing, grazing and gathering activities. Administration and technical service personnel now monitor cattle crossings, and police and technical staff reconnoitre the pastures to be sure no animals either enter before the appointed date, or remain after a certain day fixed each year for their departure.

Technical service staff also monitor a range of measures aimed at regulating the use of resources. These regulations cover:

- the type of equipment that can be used (methods of hunting and cutting browse, for example);
- species of flora and fauna that may be exploited and those that are protected;
- how reserves and protected areas are to be created;
- the permits that are needed for cutting fuelwood and browse;
- the fines that accrue to those who break the rules (République du Mali nd).

Foresters charge local producers for exploiting natural resources in the form of fines and permits for gathering woodfuel and browse. Though the cost of these permits is not large—between £5 and £10—fines for breaking forestry rules can be severe: three months imprisonment and payments of up to £600 (the value of four years' per capita production for an average producer in 1985), though they mostly amount to around £40 (about half an individual's yearly cash income). Foresters, who are paramilitary and armed, are widely feared as a result. While they fine individuals, it is generally the whole community which pays the fee as it is almost always far above what a single person or household can afford. Rural producers perceive these fines as taxes rather than a penalty for abusing the environment.

Rural development policy and the policing of rules governing resource use is carried out by officials who are outsiders to the delta. These officials are more concerned with their prospects for promotion in the forestry service than with the interests of the local community in the execution of their tasks. Moreover, they come to the job without prior training or knowledge. This is further emphasised by the policy of keeping senior staff (such as commandants and head foresters) in their posts for a comparatively short time of two to three years. The period a successor spends with the previous incumbent is normally very short and the written information left behind is often scant. The people officially the most empowered to take decisions affecting allocation of resources are therefore often unaware of local events, have little means of interpreting them, and are sent somewhere else at the moment when they begin to understand the area they administer.

Of great consequence for the management of the pastoral resources of the delta is that development initiatives undertaken over the past 25 years have failed to reach the majority of producers on the ground, and that fiscal policies have drained resources from the area.

At the forefront of the government's rural development policy in the Mopti Region were the ODR. These were set up between 1972 and 1975, largely in response to drought in the early 1970s. They were sectorally based on agriculture, fisheries and livestock, and between 1970 and 1985 were responsible for some US$75m worth of investment, nearly 90 per cent of which was provided by foreign aid (IUCN 1989).

This considerable sum was invested in infrastructure (roads, buildings), the provision of recurrent expenditure (salaries, maintenance funds, transport and so forth), and directly in productive activities (for example, capital works, wells, irrigation schemes, equipment). Under half reached this latter sector. What is more, the greater part of productive investment was in fixed capital works—for instance, the rice polders built by Opération Riz—which had been removed from the ownership and management of local producers (including herders), and in some cases was distributed to government officials. Investments were made, above all, near urban centres: the capital of the region, Mopti, or in Ténénkou on the western border of the delta. For the greater part of the rural population who live outside these areas the ODR were merely government schemes which offered them little benefit. Indeed, for many herders the livestock agency managed state access in as much as their agents were concerned in setting dates for the livestock crossings in the delta each year. This was a cost to be borne by the producers as this ODR levied fees on each herd taking part in a crossing.

Until 1991, the collection of taxes and other charges was of paramount importance for the administration. In the late 1980s efforts intensified to increase income as the government's revenues from export crops at a national level, especially cotton, fell significantly as a result of declining world prices. Until the mid-1980s, 90 per cent of the Malian budget was spent on education and the civil service, leaving 10 per cent to pay for the maintenance of infrastructure and to contribute towards development initiatives that were, for the most part, funded by foreign aid. Dwindling revenue from the export sector and rising costs of imports in the 1980s saw the government facing increasing difficulties in meeting its wage bill, leading in turn to a renewed effort to collect taxes from the rural population.

All Malians of tax-paying age (15—50 years for women, 17—55 years for men) had to pay the 'minimum fiscal' to which were added obligatory contributions to the governing political party, the parent—teachers' association, and the cooperative movement, exceptional contributions for particular projects (such as a war with Burkina Faso), and taxes on boats, guns and livestock. The bulk of these charges fell on rich and poor alike. The relentless pursuit of tax revenue became symbolic of the administration's policy for rural producers in the delta. Over 70 per cent of taxes gathered went on payment of civil service salaries, and almost all government investment was concentrated in services and infrastructure in the *cercle* and *arrondissement* towns. Most rural producers considered payment of taxes and other charges, ostensibly for development, as in fact a net gift to the administration with no prospect of benefit to themselves.

Between 1970 and 1985 the proportion of revenue from fines rose from 50 per cent of all revenue for the Forestry Department in the delta to 65 per cent, while at the same time overall revenue rose by a factor of seven until the arrival of the 1982—87 drought. Even then it was double the level it had been in the previous great drought of 1972—73, despite the fact that the later drought was far more serious for local producers.

Detailed information from one *cercle* in the delta particularly noted for its pastoral resources shows that in the years 1984—85, when the drought was at its worst, 60 per cent of revenue for the forestry service came from fines. Three-quarters of overall receipts were from penalties and permits issued to rural producers exploiting forest resources. Moreover, the data reveal that peaks in revenue coincided with the arrival and departure of livestock in the delta at the end of the wet season and at the beginning of the rains. Small stock owners, who are fined for cutting trees for forage, were disproportionately penalised for their exploitation of natural resources at times of the year (the end of the dry season) when they were most in need of cash to buy food.

Up to the early 1990s there was an inbuilt incentive for foresters to fine as a proportion of the penalty accrued to them personally. In 1984—85, five per cent went to the regional director of the Forestry Agency, five per cent to the head forester in each *cercle*, and 15 per cent to the foresters that actually imposed the penalty. Frequently the foresters on patrol would share a part of the 15 per cent with informants who came forward with evidence of a breach of the rules. Almost none of the remaining revenue remained at the local level after these deductions had been made. Instead this was transferred to the National Forestry Fund from where it was often drained into the national exchequer to meet short-term national administrative needs.

Research has shown that if all investment—that is, in the productive, infrastructural and service sectors, and including the value of labour contributed by local people—is aggregated over the 1970—85 period, its value is greater than what was extracted from the Mopti Region in the form of taxes, fines and other payments. If, however, only those investments which were made in the productive sector are included, then the region has been a net loser (IUCN 1989). Most rural people did not benefit from the infrastructure and services put in place by the ODR and the administration over this time, as these were sited mainly in urban centres; nor have they benefited from investments in the productive sector as these have been largely unsuccessful.

PRINCIPAL AREAS OF ALIENATION AND CONFLICT

As resources have become more valuable in the delta, so access to them has been thrown open to people who had no such access rights before. Those access rights have not been assured for the future by the consistent application of an effective, longterm government policy, and exclusive rights to resources have not been granted to specific groups of rural producers upon which such assurance might be based. Conditions in the delta might be described as those of 'structural chaos' in which access to resources for herders depends on a set of changing alliances with either the post-colonial state and traditional managers, or a combination of these.

In effect, customary systems function despite an array of state institutions which often compete in allocating access to natural resources. This is partly because of the seasonality of state intervention in the rural economy, which concentrates on the moment of the year (the falling-water and the dry seasons) when the delta's resources are most economically productive, giving customary management regimes a free hand at other times. In many rural producers' eyes these systems still represent a clear attribution of resources according to the widely accepted principle of the right of the first comer to manage. They also continue to function by having grafted themselves onto the various elements of the state structure and learning to play off their competing interests.

The principal areas of conflict in the delta concern the alienation of Fulbé animals to the ownership of interests outside the area. Increasingly dry conditions and the monetisation of the delta economy have broken down the barter arrangements between herders and farming communities and have obliged them to sell animals in order to raise cash to buy grain. These animals have largely been acquired by wealthy groups of non-herders: senior civil servants and merchants who contract the Fulbé to manage the herds for them (Turner 1992). Traditional contracts that existed between the Fulbé and outsiders, which were based on payments in kind (milk and the gift of young animals to the herder), are now giving way to payments in cash, with the effect that herders are less interested in taking care of the animals and managing the herds.

Conflict in the delta occurs both between herders and non-herders, and between groups within the herding community itself. Disputes between herders and non-herders principally concern farmers and pastoralists. As the delta has become drier so farmers in the area have begun to cultivate deeper parts of the floodplain that formerly contained pasture, and on the borders of the delta they cultivate the stock routes leading into the floodplains. This leads to conflict between the two groups over which areas of land should be put into cultivation, and over damage to crops by livestock. Farmers involved in these disputes often used to have strong linkages with the Fulbé, to whom they customarily gave their cattle to herd for them during the rainy season. The drought of 1982—87 killed many of these animals or obliged farmers to sell them, and in recent years farmers have invested increasingly in more drought resistant small stock which they herd themselves. This has broken down the customary alliances that used to exist between these two production systems.

More important is the conflict between delta Fulbé and outsiders, and in particular the Kel Tamasheq, who come into the area each year from the north and east. Long-standing enmity separates these two groups, and up until the 1930s the Tamasheq used to come into the delta as armed herding groups.

The Tamasheq, who rely more on dryland pastures, and raise goats and camels as well as cattle, have a greater interest in forest fodder resources, while the Fulbé, who are specialised cattle raisers, are primarily concerned with floodplain pasture. Open conflict can thus easily break out between these groups, generally over the timing of their arrival on the floodplains. The Tamasheq used to wait several days after the Fulbé had moved into a herding territory before being let in themselves, but with the nationalisation of resources in 1959, some Tamasheq groups endeavoured to force access onto the floodplains at the same time as the delta Fulbé. In 1972 an armed confrontation left several Fulbé and Tamasheq dead in the northeastern pastures around the lakes.

In endeavouring to force access to the Fulbé pastures, the Tamasheq used a strategy adopted by other 'stranger' groups to the delta—making alliances with state structures (the local administration), often through informal payments, to achieve access. Strangers ally with the state in claiming access for all Malian citizens, while local people cleave to their tradition of being the customary owners of the resources and their habitual users. It was significant in the 1972 confrontation that one of the local administrators (and the conflict took place on the frontier between two *cercle*s) was a Kel Tamasheq.

Since 1989 this conflict has taken on an altogether larger dimension with the outbreak of the Kel Tamasheq rebellion against the Malian state in the late 1980s. This has combined with the political uncertainty following the overthrow of the military regime of Moussa Traoré in March 1991, effectively to bar the Fulbé from transhumant routes to the north. According to the most recent reports, this has meant that many more animals have remained in the delta over the past four to five years, and that many animals now move to the south on transhumance each year. In doing this they traverse land that is much more heavily cultivated than the southern reaches of the Sahara, where they customarily go. The effect of this change on levels of conflict in the zone is not known.

PROCEDURES FOR CONFLICT RESOLUTION

Most conflicts are handled at *arrondissement* and *cercle* level in the delta, although serious cases can go to the regional (governor) and national level. The technical agencies (the Forestry Department and Livestock Service) generally make a preliminary report, and if violence or criminal damage is involved the police are called in. While due judicial process exists on paper for the arbitration of conflict, most cases are handled either directly by the technical agencies or by the *chef d'arrondissement* or the *commandant du cercle*. Both the technical agencies and the local administration have sweeping powers to arrest, confiscate and fine, and it is rare for the more common forms of conflict to go beyond this level. The judiciary only comes to the *cercle*s within the delta at infrequent intervals.

Two primary institutions for managing conflict are the committees (*arrondissement* level) and councils (*cercle* level) for livestock matters. These bring together the administration, the political party(ies), the technical services and producer representatives. Again, however, they rarely meet. What is more, while they include representatives of herders (often the traditional managers), the herders' ability to affect the decisions of these bodies is constrained by:

- the explicit opposition of the government to what it terms 'feudal' authority;
- paucity of numbers on the committees;
- the dominance of civil servants and technical agents;
- the fact that few herders speak French.

In general, conflict between farmers and herders over damage to crops is settled at the level of the herding leader (who might be the *dioro* or the sub-*dioro*) and the leaders of the farming community. With more serious cases of conflict, especially between delta Fulbé and the Kel Tamasheq, a range of officials are brought in—from the technical services on the ground, the local administrators, the police and the governor, and not unusually officials from the capital as well.

Settlement of disputes often seems to be carried out on an ad hoc basis, depending on who is brought in and the connections of the different parties to the dispute with people in the civil service, the political parties and the government. Strong linkages exist between delta Fulbé and national ministries in the capital as a result of the considerable wealth embodied in the herds, some of which are owned by civil servants themselves. As a rough estimate, the million or so cattle that pasture in the delta each year in the dry season—the larger part of which are herded by the Fulbé - were worth something in the region of US$85—100m at 1986 prices.

Pastoralists are not well represented in local and national decision-making bodies. The legal system provides a very uncertain and ambiguous framework of rights, and resource law is rarely understood by herders. Where a livestock parastatal (Opération pour le Développement de l'Elevage dans la Region de Mopti—ODEM) has endeavoured to set up pastoral organisations in areas neighbouring the delta as part of a World Bank project, they have met with little success. This is because of a failure by government to deal with land and water rights issues, weak and untrained leadership within the organisations, the inability of the organisations to collect revenue, and resentment towards the initiative because of unmet expectations (Shanmugaratnam et al 1992).

ENVIRONMENTAL IMPLICATIONS

It is common for development workers to argue that herders are degrading the resources of the delta, and that the time and quantity of animals that they keep are beyond its 'carrying capacity'. The notion of high risk pastoral resources in sub-Saharan Africa having such a 'capacity' is now coming under increasing critique (Behnke and Scoones 1991). For the delta it is particularly hard to define as data for longterm degradation of natural resources are scant. It is very difficult to determine the overall productivity of resources in the delta because conditions change so much from year to year, and between different localities. The technologies used to exploit the area (species mix of herds) are numerous, and in spite of many attemps to estimate a carrying capacity, no reliable figures have been reached so far.

Ways in which pastoral resources are being badly used, however, can be identified, which may mean that they could be degraded in the longer term. It is almost always the case, however, that outside influences have obliged herders to use the pastures in these ways.

In particular, it is a moot point whether the policy of keeping the animals out of the pastures until the water has withdrawn in the northern reaches of the delta—a policy first implemented by the colonial authorities, and then adopted by successive Malian governments after independence—is good for the resource. Bourgou (an aquatic grass (*Echinochloa stagnina*) which is an important dryland season fodder) regenerates best from its stems rather than by seeding. When the animals used to enter the floodplains under the customary system the pastures were still partially flooded so that the hooves of the animals implanted many of the stems into the mud, so conserving pasture (in that they could not graze it all) and providing the circumstances for its regeneration at the same time. It has also been argued that the current setting of dates for entry into all pastures has made herders concentrate in the waiting zone adjacent to the upstream end of the delta early, thereby degrading forage resources in that zone (CIPEA 1983).

It can also be argued that the government's fiscal policies are directly contributing to unsustainable use of pastoral resources through raising rural producers' cash needs: dry conditions and the breakdown in barter agreements have combined with these policies to oblige herders to exploit some pastoral resources directly for sale. In particular, farmers in the delta in recent years have taken to harvesting bourgou for sale (often pulling the plants up by the roots) and farmers and farmer-herders have invested heavily in goats, which has brought increased pressure to bear on the forest resources of the zone.

Because herders are fined specifically for cutting tree branches to provide fodder for their goats, and in general have to pay higher fiscal dues, their cash needs are raised, and the only means of raising this cash is to pursue the same strategies they are taxed and fined for. Herders are therefore obliged to continue acting in ways which the government deems destructive of the environment in order to pay for the charges the government levies to protect the environment. This amounts to the mutually reinforcing processes leading to environmental degradation described by Blaikie and Brookfield (1987) in which a higher rate of use of a resource leads to its degradation, prompting in turn a yet higher level of exploitation.

CONCLUSIONS AND RECOMMENDATIONS

The delta Fulbé developed a sophisticated and effective system for allocating access to floodplain pasture which arbitrated the interest of over 30 pasturing territories, as well as those of outside herding groups visiting the area each year. Colonial and post-colonial resource tenure law, combined with fiscal and development policies, have undermined this system without providing an effective alternative. Increasing levels of conflict, both between herders and non-herders, as well as among the herding groups themselves, now characterise the area. Mounting cash requirements and fixed entry dates into the delta are probably contributing to the degradation of the area's pastoral resources.

As a result, the delta's pastoral resources have been thrown open to outsiders who can claim rights to use the area that are upheld by the Malian state, but which the state can neither manage nor control. These outsiders often enjoy preferential ties with state structures through kinship or other economic ties, and they often do not depend on their livestock for their livelihood. Where access is available to all-comers, and the customary management systems become more and more debilitated, conditions are ripe for the 'tragedy of the commons' (Hardin 1968).

In the past six years the totalitarian government of Moussa Traoré has been overthrown, to be replaced with an elected administration. These changes have been accompanied by a growing policy rhetoric for the decentralisation of central government power and increased responsibility for local communities to manage their own development. This has provoked a continuing debate on how this might be done, in circumstances where the resources available to government (aid flows and fiscal dues) are falling.

1. Management systems and tenure policies for the delta's pastoral resources will require reform in the structure of state and local institutions, improvements of resource tenure law, and a radical shift in the direction of rural development policy and practice. To be effective they will require that management and tenure systems become more compatible with the physical characteristics of the resources they manage. In order to do this, they will need to:

— link knowledge of the characteristics of resources and dependency on the resource for the provision of livelihoods to the power and responsibility to manage;
— provide the flexibility necessary to respond to variability in natural conditions, through promoting linkages between production systems using the delta.

2. As an initial step it is important to identify local institutions for the management of natural resources among communities in the delta, not only customary institutions that existed in the past, such as the *dioro*, but effective latter-day structures. Because different kinds of producers often cohabit in the same community in the delta (that is, farmers and herders), more than one set of management institutions may exist. These institutions will need to be brought together to form one organisation representing user groups' interests. These interests should not be confined to just the inhabitants of the community, but should also include 'stranger' groups who visit the area each year.
3. The boundaries of the resources inhabitants of the community customarily

exploit, and in the past 'owned', will have to be identified and surveyed. This will allow a defined territory for communities to be established whose ownership can be vested in the management institution identified earlier, and which will have clear rights to allow or deny entry to rural producers, supported by the administration. These institutions will also need to be empowered to charge fees for outsiders wishing to use the area.

These community-level institutions will form the building blocks of wider management structures in the delta and could be the focus of rural development initiatives. A coordinating institution could exist at the level of the *arrondissement*, representatives of *arrondissement* institutions could make up *cercle*-level bodies, culminating in a regional level organisation which could coordinate and plan development strategy.

In many respects the framework for such a development structure exists already in the form of the *comités* and *conseils de développement* in the area, which have a written remit to coordinate grassroots initiatives. The crucial differences between the proposed structure and that already in place would be that effective ownership of resources would be in the hands of the communities themselves rather than the ambiguous tenure of the state, and that representation on community, local and district (regional) bodies would be by local producers rather than in the hands of technical and administration personnel.

4. A principal aim of state policy and practice in the future should be to reorient its relationship with pastoralists so as to support rather than hinder them in the management of resources. In addition to the two essential preconditions for this—the establishment of community-level management units with sufficient power to operate effectively, supported by legal title to the resources they exploit—there is a further urgent need to convert the top-down ideology and practice of state planning to a participatory approach. The technical services need to be stripped of their taxing and fining powers (these functions being taken over by the judiciary) and reoriented towards providing extension services, equipping local management committees with technical knowledge and organisational skills.

These ideas have already been mooted in documents assessing forestry policy in Mali (République du Mali et Confédération Suisse 1987), which suggest that foresters be paid in part according to the success they have in promoting conservation practices. This might be complemented by reinvesting the sums extracted from rural producers for breaking resource use laws in the area, so as to demonstrate that forestry, hunting and pastoral laws are tangibly geared toward the conservation and regeneration of the area's resources on which rural producers depend, rather than furnishing the coffers of the state and the pockets of technical service personnel.

5. The state should provide further a set of collective goods of practical use to rural producers. At a delta-wide level, the government needs to elucidate a policy with regard to the management of all the floodwater that comes into the area, specifically concerning the amount that, in the future, is going to be allocated to irrigation schemes (for instance, the Office du Niger) upstream of the delta. A further collective good that could be provided at low cost is information regarding the productivity of resources in different years, which would allow rural producers to make more efficient use of what was available.

The proposals presented here for structural and land tenure reform, and for rural development initiatives, argue that a set of interlocking measures have to be put in place in the delta in order for a virtuous circle of resource management to come into being. This circle starts with the identification of the shortterm needs of rural producers and an understanding of how producers act to meet them. This both defines measures that can be taken to alleviate those needs and identifies the resources that are most at risk. Interventions aimed at meeting short-term needs can then be directly linked to measures for the regeneration of the natural resources rural producers depend on. The success of those measures will increase the productivity of natural resources. Eventually, this would enable rural producers sustainably to meet their shortterm requirements from the production of the natural assets they exploit.

The rhetoric of recent government development policy points to a growing awareness in political and administrative circles of the need for local producers to be made responsible for the management of natural resources, on whose productivity in the end the larger part of their fiscal revenue depends. As long, however, as entrenched power remains in the hands of an urban-based elite, which for historical and political reasons has ignored the particular conditions that are found in the delta today, which believes in the imposition of 'progressive' solutions on rural producers, and which is obliged to extract as much revenue from the area as it can, the prospects for the sustainable management of the area's resources remain bleak.

4

MAURITANIA

Mohamed Ould Zeidane
Prime Minister's Office

Map 2 The Islamic Republic of Mauritania

INTRODUCTION

The Islamic Republic of Mauritania (Map 9) is a Sahelian country in which animal husbandry plays a dominant role. Pastoral production accounts for an estimated 20 per cent of gross domestic product and more than half the population depends at least partially on income from livestock and animal by-products.

Mauritania's pastoral heritage is sustained through its culture and social customs. Despite drought, 'desertification', and numerous other obstacles, nomadism and particularly the practice of transhumance are still widespread. In recent decades, however, the rural exodus has resulted in the dispossession of hundreds of pastoral households.

Pastoral production systems have undergone profound changes, as shrinking resources have stimulated many herders to take on secondary activities, such as agriculture and trading. Degradation of the rangeland and the constraints of a semi-nomadic and nomadic lifestyle have led large groups of people to set up 'new villages'. These newly sedentary people have had to learn to adapt to this different way of life.

After the 1973 drought a new class of livestock owners emerged as traders and government officials invested in livestock (cattle, small ruminants, camels) and operated a more commercially oriented system.

The pastoral land tenure system is very complex and based on various forms of law: traditional exclusive rights, *Sharia* (Koranic law), and modern land reforms. Land used for pastoral purposes is mainly in the public domain and access is unrestricted (grazing and other pastoral resources are free), although some groups do have some rights over areas they regard as belonging to them. The lack of legal clarity over the nature of pastoral land rights has given rise to various conflicts between sedentary and transhumant people, and between farmers and pastoralists.

The history of land tenure in Mauritania can be divided into four major phases linked to the political, economic and social conditions of the country:

1. The pre-colonial period, characterised by tribal control over larger or smaller portions of the national territory, which corresponded to the spheres of influence of the different Emirs.

    During this period frontiers were poorly defined and land conflicts often occurred, giving rise to numerous raids and livestock thefts. The warrior tribes (Hassane) imposed order by force of arms, shifting alliances, and resorting to Koranic law. However, this did not amount to ownership of pastoral land.
2. The colonial administration disturbed this order to serve its own interests. Unable to master a situation they did not understand, the colonial administration set three main objectives for pastoral people:

— Supplying meat to the groundnut-growing areas of Senegal, particularly by requisitioning livestock.
— Settling disputes in favour of tribes allied to France, or at least those who

remained outside the rebel movements. In this way many decrees giving rights over land and water points were issued in favour of customary chiefs loyal to the French. Even today the descendants of these traditional chiefs consider that the land concerned belongs to them.

— Fostering the emergence of individualistic tendencies, safeguarding property transactions, and opening the way for capitalist penetration and the cash economy. Privatisation of land rights has allowed the beneficiaries to use land as security for credit.

3. Attaining independence did not bring about significant changes. Law No 60.139 of 2 August 1960 mainly stressed the need for regulations based on customary land rights. In fact the state, still in its infancy, could not risk a head-on collision with popular feeling which remained inflexible and attached to tribal values. Despite declarations of intent favouring the agricultural sector, government policy in the first decade of independence was mainly directed towards the mining sector, exports from which provided the country with a positive trade balance. It was not until the 1970s, with the advent of catastrophic drought, massive rural exodus, and the disintegration of traditional methods of production, that the state began to give any real priority to the agricultural sector. Construction of the Diama and Manantali dams on the Senegal river provided the real stimulus for the public authorities to undertake land reform to increase agricultural production, open up the sector to private investors, and introduce technical innovations, thus guaranteeing income stability for rural communities, and thereby limiting rural exodus.

POLICY AND LEGAL BACKGROUND

Decree No 83.127 of 5 June 1983, which reorganised land and property rights, and its enforcement order No 90.020 of 3 January 1990, are intended to give the state prerogatives with regard to land. Articles 1—3 of the decree are quite explicit:

Land belongs to the nation and any Mauritanian, without any kind of discrimination, may, in accordance with the law, become a part owner...The state recognises and guarantees private land ownership which must, in accordance with the Sharia, contribute towards the country's economic and social development...The traditional landholding system is abolished.

State control over land was designed to achieve the following objectives:

1. to open the way for private investors, with the aim of stepping up the implementation of irrigation schemes, especially after the completion of dams on the Senegal river. This gave land to owners with the means to take advantage of new technology, to help meet national production needs and reduce the food deficit;
2. to reduce social inequalities by distributing land to the most needy social strata and specifically to foster land ownership amongst freed slaves;
3. to create employment in the agricultural sector and decentralise economic activities in order to reduce migration to towns.

In essence, this decree and its enforcement order are based on the *Sharia*. This recognises the right to private property if land is 'made productive' and if third party rights are respected. However, exclusive rights over pastoral resources are not recognised, even for people who have effected improvements or sunk wells. They merely enjoy priority user rights. In attempting to reconcile the traditional land tenure system with present day socioeconomic requirements, this decree did at least partially clarify the former situation where laws had been drawn from many different, often contradictory sources.

Development Policy

Little is known about animal husbandry, although it is the main rural activity. Pastoralists and herders have remained on the fringes of the country's economic life, and when regional development plans were devised the needs of the nomads and transhumant herders were never taken into account. Consequently, misconceptions about pastoralism are legion among the public authorities as well as the rest of the population not directly involved in animal husbandry.

The government and donors have tended to think that pastoralist production 'happened of its own accord'. All that was needed was to protect livestock against rinderpest (a very contagious cattle virus for which an effective vaccine is available) and contagious lung diseases. Current policy in the livestock sector is seeking to correct this view, enlarging the field of public intervention to cover all aspects of production (for instance, credit, pastoral management, herder organisations). Yet prejudice still remains. Although animal husbandry represents around 80 per cent of the rural sector's contribution to GDP, the investment it receives accounts for less than 10 per cent of the rural sector's budget.

Furthermore, the authorities usually think that a transhumant or nomadic herder can only receive assistance if he settles in one place. And in the event of drought or disaster, only sedentary people benefit from emergency relief. The inappropriateness of assistance granted to pastoralists was probably at the root of the massive exodus which occurred after the droughts of 1968, 1973 and 1984.

The same sort of misunderstandings permeate perceptions of pastoral culture and economy. Production is in no way assumed to be based on an economic rationale. It is thought that herders seek only to perpetuate their livestock, attempting to increase herd size at the expense of the ecological and environmental consequences (failure to destock equalling increased livestock numbers, equalling degradation of a fragile ecosystem, and so on).

Despite the fact that some herders do allow livestock to overgraze, it is often forgotten that pastoralists are the first to suffer from the destruction of grazing land and that they are therefore concerned to preserve the natural environment. By way of example, among 14 pastoral associations (PAs) set up by the government in 1988 and 1989, no bush fires were reported thanks to the efforts of the herders, who had established surveillance brigades.

The failure of government to understand the pastoralists' relationship with the environment is illustrated by the draconian measures taken to penalise them for traditional land use practices. Forestry guards are particularly hard on herders caught trimming tree branches or merely suspected of having done so. They are obliged to pay an arbitrarily calculated and immediately extorted 'hatchet tax', and if the herder does not have the necessary money, livestock is requisitioned. Mauritanians herding livestock in Mali are the main victims of this practice.

As a result of these assumptions and the fact that the country is self-sufficient in meat, the government offers more incentives to farmers, for example, in the form of guaranteed minimum prices and input distribution than to pastoralists.

Although the state has abolished the traditional landholding system, there are some safeguards for communal lands. Apart from estates devoted to intensive farming (irrigated crops) and regions where disputes between tribal or ethnic groups persist, village and pastoral communities can enjoy collective rights to land if they have contributed to the sustainability of the scheme or developed the land they claim (Article 6 of the decree covering land and property reorganisation).

The law seeks above all to encourage individual land ownership as an incentive for farmers and investors. Only cooperative organisations may enjoy rights over commonly held lands or estates. Any community which expresses a wish to retain undivided land may organize itself into a regularly constituted cooperative whose members have equal rights and duties. The same goes for any community whose land cannot be divided amongst individuals for economic or social reasons. Such reasons must be endorsed by the governor. (Article 15, Enforcement Order No 90.020, 31.1.1990 of Order No 83.127 on land tenure reform).

The establishment of cooperative pastoral associations, which began in 1988, is based on this regulatory provision.

Decree No 83.127 of 5 June 1983, which governs the reorganisation of land and property, and its enforcement order No 90020 of 31 January 1990, are not sufficiently explicit as far as land used for pastoral purposes is concerned, despite the fact that this belongs almost exclusively to the state. Both texts, which may be used in the context of animal husbandry, mainly govern land used for agricultural purposes (largely in the Senegal river valley).

Article 24 of Decree No 171 of 15 December 1982 established the forestry code and recognised the user rights of rural communities in respect of forest resources (use of pasture for domestic animals), giving priority to local people (Article 26). Rights to grazing in classified areas—that is, those subject to a classification order—are more restricted: herders must not put in any installations, even temporary, let alone use tools to cut vegetation (Article 32). Only the trimming of small branches of non-protected species is authorised.

These two decrees are virtually the only official texts dealing with pastoral land tenure. Improving rangeland development, nevertheless, requires legal provisions and the granting of more tangible rights to legal entities and individuals wishing to exploit the rangeland. It is true, however, that it would not be advisable to attempt to establish a code now while animal husbandry is undergoing radical changes, except perhaps in clearly defined regions of the country.

Livestock development policies have dealt exclusively with animal healthcare and the establishment of pastoral water points. These measures have established favourable conditions for an increase in livestock numbers while the country's fodder supply remains limited, and have contributed to the overuse and sometimes degradation of pasture. The rangelands and the grazing they support currently represent 'a scarce resource which does have economic value' (World Bank 1986). However, access to these resources (grazing and water points) has almost always been free, with the exception of some periods when Hassane tribes exacted tribute in respect of their domains.

The wish of certain people to maintain rights therefore contradicts that of others who seek to pursue self-interest through the assurance of free access to grazing land throughout the country. This contradiction has been 'thrown into sharp relief by drought, but is not the consequence of it' (Bonte 1987).

The government was prompted to establish PAs, initially on an experimental basis, by the need to eliminate the degradation and overexploitation of rangeland. The principle was to give pastoralists the right to take over the management of their area in exchange for a commitment to maintain and develop the land. The teams responsible for setting up the PAs included experts from many disciplines (sociologists, veterinarians, cartographers). The identification of a PA, the preliminary phase, was made through contacts and surveys among groups of herders located within a given area. Consideration was given to historical, economic, sociological, climatic and other factors.

PAs are to be given:

- *The opportunity at their request to lease areas of pasture to which they may, on granting of the lease, reserve the right of entry. The same goes for water points provided by the state for pastoral use;*
- *The function of auxiliaries of the public authorities in terms of rural policing (combatting bush fires, tree felling etc).* (Circular No 19, Ministry of the Interior, 14 August 1990)

The internal regulations of the PAs, which must be endorsed by the administrative authorities, are supposed to define the rights to be granted to third parties. Indeed, to avoid disputes with groups of transhumant or nomadic herders making seasonal use of the territory belonging to the PA, there needs to be a consensus to protect the rights of these visiting groups. They could pay a fee, for example, bearing in mind that the PA association is responsible for renovating wells and improving the rangeland.

To help the new PAs to attain their objectives, various technical services have been provided within the Livestock Department (the supervisory body). For example, the pastoral water supply service is responsible for renovating and installing water points for the benefit of the PAs. In return, the PAs must provide assurances that they will maintain these wells and respect the rights of third parties.

Funding through the financial service of the Livestock Department is available to improve production potential. Animal fattening or dairy production projects, including the production of fodder crops, can be funded with a contribution in kind and/or cash from the members of the PA. The pastoral management service not only deals with the regeneration and introduction of fodder plants, but also provides technical advice and training to the herders on such diverse topics as combating bush fires, establishing fire breaks, and forbidding access to certain areas. Each PA elects its own pastoral management committee whose duties are to:

- rationalise pastoral land use;
- protect the rangeland;
- improve pastoral production;
- evaluate and monitor the situation.

TRENDS IN PASTORAL DEVELOPMENT AND WELFARE

Current policy is for the state to monopolise practically all pastoral land, leasing parts of it to herders organised into PAs which undertake to manage it properly. However, pastoral people distrust the state and although traditional landholding arrangements have disappeared in many cases it has proved difficult to impose new ones. Extensive herding practices and the sedentarisation of nomadic and semi-nomadic pastoralists in 'new villages' militate against any definitive redistribution of land rights.

Until the full application of the policies laid down by the government occurs, access to pasture remains free almost everywhere, which complicates the task of the PAs. How can they be expected to improve the management of rangeland when other pastoralists from elsewhere continue to regard land as free and available to them? An already serious situation has been exacerbated by the diminishing scope for transhumance since the closure of the Senegalese border following clashes between Mauritania and Senegal, and since the Malian authorities imposed ever harsher conditions on access for livestock from Mauritania.

However, pastoralists are now increasingly aware of the need to combine their efforts to preserve the environment and form an important pressure group which can influence political decision-making. By organising, they also become creditworthy for banks, and present a viable focus for support from government and international institutions. Given time, they should be able to reach a new equilibrium that takes into account environmental considerations as well as their own aspirations.

The profound upheaval in land tenure arrangements stems from three contradictory causes:

1. The presence of collective interests symbolized by tribal units, and invoking ancestral rights over land and pastoral resources.
2. The existence of opposing forces represented by mobile groups, but also by new livestock keepers—that is, that class of owners comprising retired high officials and merchants who purchased a number of livestock during the drought years at very low prices as a means of accumulating capital. These groups are particularly anxious to institute free access to pasture, declaring that this is the only possible option in the face of successive droughts. It is estimated that 40 per cent of cattle belong to these new livestock owners. They also hold a higher proportion of the camel stock, but appreciably less small ruminants than traditional herders (World Bank 1986). Some religious movements also consider collective rights to be superseded by the rights of

the whole community of believers to use the rangeland as they see fit (Bonte 1987), implying free access for anyone owning a herd.
3. The political will of the state is in conflict with the preceding two landholding systems. The state denies any tribal ascendancy over the land and wishes to cede rights only to PAs whose objectives are to improve and manage the rangeland. At the same time it is opposed to absolute free access as this leads to anarchy and the waste of scarce pastoral resources.

Apart from desert areas, pastoralism is present everywhere in Mauritania—testimony to a rich heritage when animal husbandry was the country's only economic activity. While camels used to be restricted to the north to avoid sleeping sickness and other waterborne diseases, large numbers of them have infiltrated the south over the past few years.

Livestock Development Projects

Livestock development projects came into being after the great 1968—1973 drought at a time when the herders were at their most vulnerable. Apart from activities undertaken by the government, development support has also come from bilateral donors, international or regional organisations, and NGOs.

The World Bank has been operating in Mauritania since 1986 under the aegis of the Second Livestock Development Project for Mauritania (Livestock II). This project has an overall budget of more than US$18m financed through the International Development Association, in collaboration with the African Development Bank, the Organisation of Petroleum Exporting Countries, and the Nouakchott Municipality. The project covers a series of activities, the most important components of which are the: establishment of PAs; study and monitoring of production systems; animal production; and training.

Through these activities, the government hopes to put an end to overexploitation of rangeland and restore the balance between the numbers of livestock and fodder potential. If there is to be any chance of success, pastoralists and nomads must be involved in decisions that affect them. This is why the activities of the various components are experimental and will not be maintained or extended to other areas unless their impact is deemed to be positive and unless the herders agree.

Fifteen PAs were initially established and the government then decided to extend the experiment to the rest of the country. The process of creating new PAs stopped in 1992 after the constitution of 39 associations, and activities since then have moved towards the consolidation of existing PAs. A project for natural resource management is now under discussion with the World Bank. It is not clear whether this project will lead to the creation of further PAs or not.

Nomadic pastoral associations (NPAs) are for herders located in areas of the country which are generally unsuitable for settlement and only visited sporadically by groups of pastoralists. NPAs are not based on the same principle as PAs: 'The territorial aspect tends to take a back seat here in favour of a unit based on a group of people traditionally using a grazing area by virtue of a series of wells.' (Bonte 1990).

There is practically no water and rainfall is extremely variable in these areas of the country. The only way to use the resources, which can sometimes be substantial, is to move back and forth from one well to another according to the pattern of rainfall. This means that contact with these groups occurs only when the camp is next to a well. Groups of people rearing camels, who are still attached to a nomadic way of life, constitute virtually autonomous units, unlike other systems found in the south of the country.

The establishment of NPAs should make it possible to gain a better understanding and management of the conflicts which often occur between these constantly moving family units and agro-pastoralists in surrounding areas.

Village pastoral associations (VPAs) correspond to the system of community herding found in the Guidimakha and Senegal river valley regions. VPAs take into account the transformation in landholding which has resulted from the development schemes in the river valley.

The European Development Fund subsidised a project (Projet de Financement pour un Projet de Développement de l'Elevage dans le Sud-Est Mauritanien) the objectives of which were improved animal health, pastoral water supply, and support to the livestock services, as well as outreach work with herders. Despite its generous funding (close to 200 million Ouguiya), the project's impact has not been substantial in pastoral regions. The most interesting component has been the establishment of pilot villages. The idea was to provide on-the-spot assistance and training and to fund simple integrated rural development activities, such as animal-drawn agriculture, market gardening and smallscale poultry keeping. Any remaining funds are used to extend the operation to other villages.

The French Aid and Cooperation fund regularly gives grants for ad hoc activities. For example, the aims of the Trarza livestock development project are as follows:

- assistance and training for herders;
- research and development for camel production;
- intensifying animal husbandry in irrigated areas.

The German GTZ agency is involved in several integrated rural development programmes, including one in Achram-Diouck. In addition, several donors and regional organisations have funded work to drill and restore pastoral wells in areas clearly suited to pastoral activities. The Arab Fund for Economic and Social Development, the Kuwaiti Development Fund, the United Nations Development Programme, and Food Security Commission, and the West African Economic Community are among the main donors. However, the objective set by the government—to establish a network of large diameter wells, six to 15km apart in the grazing areas—is far from being achieved in spite of this large programme.

There are other donors in the social and economic field (agriculture, nature conservation, health and social affairs projects). Most of these projects are concerned with formerly nomadic peoples who are increasingly settling in new places where there is no infrastructure. Government agencies are mainly involved in the agricultural field (for example, market gardening, rain-fed agriculture, small dams). Several women's cooperatives have also received organisational support and assistance, which has helped to increase production.

Apart from the World Bank funded Livestock II project, which seeks to grant collective rights to herders, all other projects have ignored land tenure issues (especially in pastoral areas). However, the decline of tribal and, generally speaking, traditional rights has created a situation in which much of the land has no effective 'master', which is bound to create obstacles for its use, whether agricultural or pastoral. The state theoretically controls vast tracts of land, but in practice its hold is tenuous. Ways and means to extend state power that respect local needs and the complexity of relationships around land issues are therefore urgently needed.

Women

Women have long been marginalised in rural society. In pastoral societies, where men are often on transhumance and concerned only with the amount of pasture and the state of their animals, women exert greater influence, although this is still not very strong. They have often been, however, the catalysts for the changes that have occurred in pastoral societies. With the drought, the emancipation of slaves, and the massive exodus of men towards the towns, women have taken on an increasingly important role as decision-makers.

Under the communal regime only men had the right to speak publicly at mixed-sex meetings and take decisions. By contrast, individual land ownership, as legislated for in the 1983 reforms, enables women to become owners. Consequently, women's cooperatives are increasingly gaining access to land and undertaking rain-fed and flood-retreat agriculture, as well as market gardening, thus introducing new dietary habits.

As marriage is not based on shared property women can hold individual property and land, acquired either by inheritance (in accordance with the *Sharia*), dowry, or gift (from a father, brother, or other relative). They may also undertake transactions in respect of such property, such as purchase, sale or gift.

Women are entitled to own cultivated fields, palm groves in the Adrar, Tagant, Assaba and Hodh Gharbi regions, and compounds or houses. These rights have almost always existed, but are being increasingly enjoyed for their own benefit. This increased awareness is evident, even among women in recently settled pastoralist groups.

In pastoral households, the woman holds her part of the herd (mainly small ruminants, but sometimes cattle and camels). However, it is the husband who is generally responsible for their management. This does not, however, give him specific rights to the animals held by his wife. Moreover, in the event of divorce, the woman takes her share of the herd back to her home, thus breaking up the family livestock unit. The frequency of divorce in recent decades has made women very protective of their economic independence. But prejudices still persist: even though their rights to property are recognised, it is always their fathers, brothers or husbands who manage the property on their behalf.

A relatively recent socioeconomic survey of about 30 pastoral women showed that their time was approximately taken up as follows:

- domestic work, 62%;
- processing animal produce, 9%;
- taking care of animals, 6%;
- agricultural work, 2%;
- marketing produce, 8%;
- personal care and leisure, 13%.

The study also revealed that 28 of these women said they had property which they usually manage quite independently, apart from the animals entrusted to their husbands. For women to assume successfully their role in society as property holders, they will require assistance and training.

PRINCIPAL CAUSES OF ALIENATION AND CONFLICT

Except perhaps in the Senegal river area, customary ownership of land, including pastoral land, still predominates. While access to pastoral resources is still largely determined by the traditional land use system, the disruption created by drought over the past two decades has obliged pastoralists to adopt different survival strategies, particularly in respect of herding and managing their livestock. For example, pastoralists now tend to gather around water points close to grazing land on order to avoid long journeys and further marginalisation in the event of another drought. They engage herdsmen to watch their livestock and now buy raw or processed by-products for animal feed.

One of the most serious problems to emerge from the 1973 drought was a substantial change in livestock ownership: many herds were transferred from pastoralists to merchants and government officials, who purchased them at very low prices. As a result former owners became herdsmen, often looking after their old herds. Since they started to accumulate livestock, the new owners have tried to gain at least partial control over pastoral resources. In particular, they have attempted to break down the traditional rangeland management systems and institute free access to grazing and water points, which they are in a position to do because they have money and political clout. To counter such a process, the government set up the PAs in order to enable traditional pastoral systems to survive by granting them usufructuary rights over traditional grazing areas, and allowing them access to credit.

Important social changes have also occured in the development of internal power relations, settlement, changes in dietary habits, new divisions of labour, loosening of traditional solidarity links, and new types of gender relations. These elements have all influenced landholding and may be the source of future conflict. Local administrative institutions, which are supposed to ensure respect for laws regulating access to land, usually only tend to arbitrate between communities from afar, relying on the customary chiefs to attend to local matters. Conflicts, then, tend to be the subject of provisional judgements in favour of one party. However, where animal husbandry is transhumant and seminomadic, the local administration cannot exert effective control over pastoral resources because of the scale of livestock movements from one region to another and the dangers this can involve (for instance, in the case of drought). The only exception to this relates to quarantine in the event of a serious epizootic disease, such as rinderpest. Then the animals are confined until vaccination has been completed and the danger has passed.

State attempts to arbitrate often consist of trying to respect the power relations between the opposing groups, both to avoid trouble and because of a lack of explicit regulations. Sometimes novel solutions are found, for instance, by leasing areas to a tribal group. Such a group, composed of semi-nomads wishing to become sedentary, would have enormous difficulties in finding a place to settle for agro-pastoral activities without a lease granting explicit provisional rights. However, intervention by public authorities is often resented by the indigenous people of the area chosen for settlement, who see themselves in danger of losing their traditional rights.

In actual fact, relationships between groups in the same agro-pastoral or pastoral region are defined by links forged through their common history. Pastoral societies who have access to the same pastoral land and water points recognise each other's rights. These used to be safeguarded by agreements based on the power relations between them, and each tribe and group knows very well what they can do based on past arrangements and current circumstances. So agreements (generally unwritten), drawn up between groups tend to reflect this situation. The disruption of such agreements between tribes, as a result of the intervention of the state and the internal dynamics of pastoral societies, has further complicated the management of rangeland already affected by land degradation and excessive livestock numbers.

Changes in power relations have also occurred within tribes which threaten their cohesion. Sometimes sections of the same tribe break away, for instance, if the customary chief decides that the group should become sedentary. There have also been instances where sections of a tribe have come into conflict with other sections of the same tribe by denying them access to resources. Furthermore, there is now a tendency for 'oppressed' groups within a tribal unit to step up pressure to gain greater access to land and other means of production. For example, the Haratine (descendants of slaves) who are sedentary agro-pastoralists are laying claim to ownership of the land they have cultivated as share croppers for generations.

Sedentarisation

The extent of land tenure problems varies from one region to another and from one production system to another. In regions with considerable agro-pastoral potential, the establishment of villages by groups of pastoralists often leads to discontent. Conflict is most serious if new arrivals want to have access to arable land. However, a distinction should be made between nomadic and semi-nomadic pastoralists. Nomads do not actually lay claim to land they have crossed during their traditional migrations. Any attempt by them to settle permanently in a given region usually arouses the hostility of local tribes who consider they have priority land ownership rights. On the other hand, semi-nomadic pastoralists do lay claim to areas they cross during transhumance. They see settlement in any given part of these areas as an absolute, indisputable right. In any event, conflicts generally arise between new arrivals and already settled groups of people who have invested in the land, and who have also dug wells over which they have priority, if not exclusive rights.

ENVIRONMENTAL IMPACTS

The Saharo-Sahelian climate of Mauritania is characterised by low and variable rainfall. Only the Guidimakha Region (with a total area of 10,000km^2) in the extreme south receives annual rainfall of around 500—600mm. The droughts of 1968, 1973 and 1984 dried out areas which used to be well-preserved. Organic nutrients in the soil have been carried away by wind and the whole area has become degraded. Despite this, animal and human pressure remains at the same level, with herding and agriculture continued as before. The desert margins have therefore become severely degraded.

Like other rural populations, pastoral groups make use of natural resources in several different ways. Tree and forest resources are, for example, used for firewood and manufacture of various implements and tools. Woodland areas crossed during nomadic migrations or transhumance are also called on to feed livestock in the dry season by the cutting of branches or even whole trees. Some pastoralists infringe legislation to safeguard their livestock: Mauritania used to have a density of around 50—100 gum trees per hectare, covering an area of about 165,000km^2. However, the gum trees are now disappearing because they are being used to feed livestock.

Overexploitation of grazing (especially on the dunes) by camel herds has led to widespread shifting of sand, thus resulting in the 'desertification' of whole regions, particularly in the Trarza Region between Nouakchott and Rosso, and between Nouakchott and Boutilimit. On occasion, pastoralists have accidentally caused bush fires. However, there are also other causes of environmental degradation which include settlement of pastoral peoples, increased charcoal burning and agricultural expansion.

Agricultural expansion takes place on the richest land and valley floors, which also contain the best pasture for livestock. Agricultural schemes strip off vegetation, making the soil vulnerable to erosion, particularly in the Trarza, Brakna and, to a lesser extent, Gorgol regions. Expansion of agricultural activities has adversely affected the fauna and flora and is associated with a greater population density in formerly uninhabited regions. The establishment of new villages further degrades the surrounding areas for several kilometres. To combat this, the government has set up an ambitious dune fixation programme. It is not just a matter of protecting settlements, but also of protecting the region within a radius of several dozen kilometres.

The introduction of irrigated crops in the Senegal river valley has reduced the country's cereal deficit, but there is a risk that the land within these schemes will become degraded as a result of mismanagement and a reduction in technical standards in the blind pursuit of profit to the exclusion of all else.

The degradation of areas around pastoral wells set up by the public authorities over the past two decades is particularly visible in Trarza. Where wells are close together degradation is more widespread because the pasture around each well has been destroyed, creating desert-like expanses. The trend towards sedentarisation compounds the problem—it is estimated that two out of three nomads have abandoned nomadism—and this is likely to continue until a new balance is found.

Furthermore, despite the introduction of gas as a source of energy and the incentives for its adoption offered to town dwellers, charcoal is used on an ever growing scale. It is estimated that the consumption of firewood is at least eight times higher than the natural growth of available forests. To meet the demand, illegal charcoal burners fell hundreds of thousands of trees every year. Whole regions of rich pasture where animals used to graze all year round have thus been totally or partially degraded. In Trarza, where a PA was set up in 1988, charcoal burners have been decimating the forests for several years to the detriment of the herders who have suffered most through lack of forage. Despite complaints to the relevant authorities by the president of the PA, these have not been followed up. This is because important financial and political interests are involved in this affair, as they are in other well watered regions of the country, such as Guidimakha, Gorgol and Brakna.

Dams and other water schemes, although important for agro-pastoral activities, can adversely affect the environment as they tend to attract people and livestock and disrupt the ecological balance of the regions concerned.

CASE MATERIAL

In order to make a proper examination of the problems associated with pastoral land tenure in Mauritania, three climatic/ecological zones must be distinguished:

- the southeastern region, mainly given over to pastoral activities;
- the Senegal river valley zone and the regions historically associated with it;
- the north, which has very low, erratic rainfall.

There seem to be few documents dealing specifically with land tenure issues in the southeast. It is, however, an area of traditionally nomadic or transhumant herding, which has experienced an unprecedented rate of sedentarisation over the past 10 years or so, usually associated with other activities such as trading or agriculture.

The valley zone has been of more interest to researchers, NGOs and funding agencies, in view of its geographical, historical, economic and social importance.

This zone was an early target for French colonisation, it now benefits from two large dams—Diama and Manantali—and it represents a crossroads for black Africans, Arab/Moorish farmers, pastoralists. Bonte (1983) has traced the development of pastoral landholding arrangements since the reign of the Emirs. Under the Emirs, such arrangements directly reflected the social and political structure of Moorish society: the Hassanes held political power, the Zewayas (*marabouts*) devoted themselves to religion and animal husbandry, and the Zenagas depended on the Hassanes. The Zewaya managed pastoral territories according to transhumance routes or water points dug by them. Areas of territorial influence could overlap without provoking conflict between tribes. The Emirs' authority resided in land rights and armed intervention in the event of open conflict between the Zewayas.

French colonialism, which met with opposition from the Hassanes, tried to reduce the power of the Emirs by distributing land to loyal groups, thus undermining traditional territorial organisation.

The current situation must be understood in relation to the internal dynamics of Moorish society, the demands made by the emancipated slaves, the Haratines, and the changes associated with the droughts since 1968 (Bonte 1987). However, the most remarkable aspect seems to be linked to access to pastoral resources, where two contradictory trends confront each other. The first, developed mainly by tribal communities, seeks to affirm collective land rights. The second evokes the principle of free access to pasture, said to be a gift from God, that nobody has the right to reserve for their own use.

In the Senegal river valley land tenure has evolved from traditional systems (rain-fed and flood-retreat cropping, animal husbandry, fishing) to a new system based on irrigation (Boutillier and Schmitz 1986). This has created difficulties for pastoralists related to land tenure and access in an area of cultural and economic interaction (between pastoralists, farmers and fishermen). Under traditional relationships, pastoralists had access to the river through transhumance corridors and could use agricultural residues by grazing the fields. These rights are now contested by farmers.

In the north of the country, land tenure problems are less acute, because grazing is limited and water is in short supply. Nevertheless, large herds of camels cross these regions from one well to another, using sparse, unreliable pastoral resources. Grasses which do grow there, however, usually have a higher nutritional value than in other, better watered regions.

In these semi-desert areas, frequently visited by camel herds belonging to new livestock owners, the role of the Emir as guarantor of security over a given territory and the Zewayas has remained entirely complementary, giving this region a cohesive and clearly defined landholding system (Bonte 1987). Generally speaking, however, few in-depth investigations of pastoral land tenure have been undertaken and the strategies of the various pastoralist groups are still little known.

Data from the Ministry of Rural Development show that in 1990 the total number of livestock in Mauritania was estimated at 1.35 million cattle, 8.5 million small ruminants, and 9.5 million camels. Almost half the population have an occupation related to pastoral production (herders, livestock dealers, collectors of hay and straw, well diggers and so on). The rural population of around 1,200,000 inhabitants makes a living mainly from animal husbandry and each household draws at least part of its income from animal production.

Consumption of red meat at national level amounts to an average of 22.5 kilograms (kg) per person, divided as follows: beef (6kg), small ruminants (6.5kg), and camels (10kg). Consumption per head has been falling from a peak of 30.5kg in 1979, equivalent to a drop of 26 per cent over 10 years (Gaye and Zeidane 1991). Rural people and nomads in particular have suffered from this reduction in protein consumption, and food shortages cause many health problems.

PROCEDURES FOR CONFLICT RESOLUTION

Generally speaking, the government does not want to get involved in local conflicts, preferring to allow collective self-regulatory mechanisms to operate. The persistence of drought, the shrinking of rangeland resources, and the emergence of a new class of absentee livestock owners have, however, forced the state to take a stance.

The most frequent conflicts arise between tribal groups and are mainly settled by consultation between traditional assemblies (known as *Jemaa*). A decision taken by a *Jemaa* is binding on the whole tribe.

Traditional pastoralists find themselves in conflict with farmers who wish to plant crops in traditional grazing areas and who prevent livestock from gaining access to water pools. Lack of intervention by the government tends to encourage farmers to lay claim to disputed land, since Koranic law recognises the rights of those who make land productive. Conflicts, when they do arise, are sometimes bloody and are usually only partially resolved.

In the Senegal river valley the state gives priority to investors in the agricultural sector. Furthermore, some think that one of the main objectives of the new land tenure law is to expand cash cropping (rice) in the valley, in order to get a return on the huge investment made by the state in dam construction. The most serious dispute concerns access to the river itself for watering, and to valley pastures. Although they live in the region, the Keur Macene herders have not been able to find corridors to the riverbank since the irrigation schemes were set up. Land surveys are currently under way to identify a satisfactory solution for all concerned—investors, local farmers and herders, and transhumant groups accustomed to seasonal migration within the valley region.

Traditional pastoralists also come into conflict with sedentary or semi-sedentary agro-pastoralists who try to control their access to pastoral resources by forbidding animal watering at a well, or marking out disproportionately large living areas into which visiting animals are not allowed.

Transhumant pastoralists mainly cross areas which either 'belong' to them or have been the subject of intertribal agreements, at least as long as they are in Mauritania. Although the decree governing land and property reorganisation stipulates that 'all wells and boreholes situated outside private property are declared to be for public use' (Article 22 of Decree 83. 127—see above, p. 76), the government does not always act in favour of transhumant herders, and allows the tribal authorities to resolve disputes.

Pastoral groups living in the same region may have different herding strategies. This is the case, for instance, in the Dienke area of the Mougataa de Tintane region to the east of the country. Here herders from the Ladoum Moorish tribe cohabit with the Peuhl, who are comparatively recent arrivals to the area. The Ladoum have forbidden the Peuhl, and other Moorish tribes, to dig wells throughout the area. According to them, making do with a small number of wells limits the access of transhumant herders to the Dienke Region. This both preserves resources and allows local people to enjoy the use of them. This intelligent strategy also has the advantage of protecting this area against certain degradation should it be totally open to transhumant and nomadic herds. However, it disadvantages local people who do not have wells and are thus often obliged to move elsewhere.

Finally, conflicts with charcoal burners tend to be settled in the latter's favour, as they enjoy strong support within the administration and are responding to demand from people who are still heavily dependent on charcoal in their daily life.

Having long remained on the fringes of the country's public and political life, pastoralists are now trying to make their voices heard. Unlike other sectors of the population, and as a result of their migrations, they are very ill-informed about current regulations and agreements made with neighbouring countries in respect of transhumance. Pastoralist nomads have largely ignored interstate borders and state legislation for the regulation of access to pastoral resources.

While they have no direct representation within official bodies, they nevertheless enjoy a degree of influence through family connections to those in high level government posts. Whenever disputes arise with other groups, they generally call on these relatives to settle the conflict. It should be remembered that conflicts over land are settled according to the norms and codes recognised by the opposing parties. The state legal system rarely intervenes decisively in inter-tribal disputes.

PAs are also responsible for preventing conflicts between pastoralists and acting as pressure groups to counterbalance other local and national powers. In addition, since the droughts the government has been aware of the vulnerability of pastoralists despite the good rainfall recorded over the past five years.

Among the stated objectives of the National Federation of Mauritanian Farmers and Herders are the promotion of both agriculture and animal husbandry, as well as assistance for rural societies to ensure they are better represented within official bodies and in dealings with public authorities. However, the federation has been established mainly by businessmen and new livestock owners, who have very little practical experience of pastoralism or its attendant problems.

The National Association of Mauritanian Herders (ANEM) describes its objectives in a similar vein, but it does not represent real pastoralists either. Furthermore, since its establishment ANEM's activities have sharply decreased. PAs are probably the most effective means of representation, primarily because they are based on the real circumstances faced by each group of herders. Officials are also local people elected on a free, majority vote by delegates from all settlements and camps. Finally, they provide material, human and technical support to the herders and grant them user rights over land. Once they have become widely established, they will probably be in a better position to resolve instances of crisis or conflict. However, PAs are still in their infancy and need more experience before they can really influence the public authorities, agencies and banks.

CONCLUSION AND RECOMMENDATIONS

Climatic transformation has had a considerable impact on the pastoral people of Mauritania, who are still unable to find a sustainable response to the many pressures they face. Nomadism is declining throughout the country, and thousands of pastoralists no longer have herds or any hope of being able to continue a pastoral existence. Their sedentarisation leads to conflicts that are almost always over land, between them and local people. In regions where agriculture is feasible, the struggle for control over land and water points can be bitter.

Tribal and traditional power is no longer what it was, although some form of ascendency over land by certain tribes is tacitly recognised by both the public authorities and other bodies. The question of access to pastoral lands is unclear and there are two opposing views: those favouring free access and those preferring controlled access by allocation of rights to pastoral groups. However, every year thousands of hectares continue to be degraded, largely as a result of human activity. Pastoralists, like other rural people, must bear some responsibility for this.

Recommendations

This complex situation calls for a series of measures, bearing in mind the following points:

1. The struggle against environmental degradation must be waged in all regions. No form of land rights, whether traditional or not, should be allowed to cause destruction of plant cover. The forestry code is quite explicit in this regard, but is not well applied in practice.
2. Traditional pastoralists who practise extensive or semi-nomadic animal husbandry have a considerable part to play in the country's economy. Care should be taken not to disrupt an existing order, which has proved its worth, in favour of modern livestock keepers who are inexperienced in animal husbandry and whose knowledge of the environment is minimal. Rather than trying to establish a formal pastoral code that would be difficult to apply to a system in a state of flux, it would be better to direct efforts towards more flexible guidelines adapted to the circumstances of a given place, that could be revised and improved from time to time.
3. Organisation into pastoral associations, as begun by the Livestock II project, should be pursued. In this way, pastoralists could wield more clout in political decision-making and gain access to credit which, along with the introduction of new techniques, is the only way to improve herders' living standards and increase productivity. User rights granted to these associations should be strengthened, while protecting the rights of third parties.
4. Smallscale herders often have quite different interests from large livestock owners. Their greater vulnerability means that they rapidly abandon pastoralism, in the event of problems, to act as herdsmen for other people, or go to towns looking for work. The government should make a particular effort to protect their interests by guaranteeing access to land and supporting livestock markets, because it is the small producers who depend exclusively on their livestock and who suffer most from a drop in prices.
5. Development planning should be based on pastoral production systems and should ensure the greatest possible participation of herders. In this way many problems would be resolved and the common misunderstandings between government and herders could be eliminated. However, the pastoral econo-

my would have to be fully integrated within the national economy, and new strategies would have to take account of the livestock sector.
6. Questions related to land tenure rights and legal issues need further discussion with pastoralists.

Because the issues are so complex, the following could usefully be done:

- conduct anthropological surveys to understand better the history of the land tenure issue, traditional tribal structures and the pattern of collective land ownership;
- conduct surveys about the current status of land tenure in the various areas;
- review all legal documents and reports dealing with land tenure.

NGO efforts should also be guided towards practical activities. With regard to land rights, NGOs should assist by working with PAs that are sufficiently active and well informed in an effort to call a halt to environmental degradation and irresponsible human activity. Livestock corridors need to be established in agricultural areas, such as the Senegal river valley. NGOs should also assist former pastoralists congregating in the towns to reconstitute their herds or to find jobs in the animal husbandry sector. Greater coordination between NGOs is essential, as their activities are often duplicated and at times even contradictory.

The threat to the future of pastoralists and pastoralism in Mauritania should prompt all institutions and agencies involved in development to combine their efforts to assist the thousands of people who are in distress, and yet whose activity is vital for the national economy.

5

SENEGAL

Pastoral Network (REPA)[1]

Amadou Tamsir Diop
Senegalese Agricultural Research Institute,
National Livestock and Veterinary Research Laboratory

Ibrahima Niang
Livestock Service

Alioune Ka
Ecological Monitoring Centre

Aboubakrim Deme
Associates in Research and Education for Development

Map 10 Senegal

INTRODUCTION

In Senegal (Map 10), 12 per cent of the population are Peuhl, the largest ethnic group after the Wolof and the Serere. Their main activity is herding and they hold more than 50 per cent of cattle, sheep and goats. Despite their demographic, economic and social importance, neither the Peuhl nor their livestock keeping system have been fully integrated into Senegalese society because livestock keeping and its associated use of the rangeland does not seem to accord with the state's production objectives.

Policies to develop and control the rangeland, accompanied by legislative and regulatory measures, have meant that pastoral peoples find it increasingly difficult to practise their traditional forms of livestock keeping. Pressures on land use from a constantly increasing population and the endemic depletion of natural resources as a result of successive droughts compound the problems.

POLICY AND LEGAL CONTEXT

Legislation

The French landholding code introduced in Senegal in 1830 favoured the type of land ownership as practised in the West. It enabled the colonial administration to take over traditional lands said to be 'vacant and ownerless' (Niang 1982).

In 1964, Law No 64—46 concerning state-administered property, and Decree No 64—573 to enforce it, ensured that the management of the landholdings of traditional owners was transferred to public authorities. Article 2 of this law provides for the nationalisation of land by the state with a view to ensuring its rational use and exploitation in accordance with development planning and programmes.

This legislation had three main objectives:

1. legal: to take account of traditional rights and colonial and Muslim law;
2. economic: to improve land use, centralise decision-making, and encourage participation of grassroots communities in the management and exploitation of land;
3. political: in view of the regime's socialist perspective, the participation of local communities in development must necessarily be preceded by their empowerment.

Four categories of land were defined: urban, pioneer, classified and country. Country areas cover land that is regularly used for rural settlements, agriculture and livestock keeping. Pioneer areas are those being developed in accordance with development and planning programmes.

In 1972, the administrative, territorial and local reforms promulgated by Law No 72—25 confirmed the state's authority over the pioneer and classified areas, and ceded the urban and country areas to the local authorities. The country areas, where grazing lands are found, were entrusted to the rural communities to be administered through a rural council, part of which is democratically elected and part designated by the cooperatives.

In 1980, norms governing the use of grazing lands were laid down by Decree No 80—268 which defined four types of rangeland: natural pasture, fallow land, improved pasture, and post-harvest grazing. This legislative measure defines how rangelands should be organised and exploited for farming:

- all clearance for farming was forbidden within areas of natural pasture and close to areas where livestock gather (boreholes, vaccination centres, etc);
- natural pasture, classified forests and livestock tracks, had to be marked out;
- a buffer zone must be left around areas where livestock gather;
- classification or declassification of any area could occur only after detailed study by specialised commissions;
- those fields authorised for herding areas must be protected by a hedge or fence against damage by animals;
- the use of pesticides or potentially dangerous chemicals was regulated.

The decree also stipulates how pastoral water points should be managed and used. It states that prior authority is necessary if uses other than pastoral or human are being considered, that their exploitation may be forbidden by the authorities whenever required, and that agricultural activities and the establishment of camps around them are prohibited.

The use of the rangelands by camel herds is regulated by Decree No 86—320 of March 1986 and the order to enforce it. The raising, ownership and use of camels is now authorised only in Dagana and Podor and the north of Louga Linguere and Matam departments.

The state gives rural communities rights to use land:

...in accordance with the ability of the beneficiaries directly or with the assistance of their families, to make the land productive in line with the programme laid down by the council. (Law No 72—25 of 1972 concerning rural communities)

The issue here is the definition of the notion of 'making productive' in accordance with the notion of development. In principle, Article 10 of Law No 72—

1288, covering the terms for granting or withdrawing the allocation of land, provides that the prefecture (local administration) should issue an order, if necessary for each rural community, setting the minimum conditions in this regard. In practice, this has been applied by the exclusion of pastoral groups.

Rural communities have considerable rights to 'develop' their own areas. These are, however, limited by further legislation. Decree No 64—573 states that 'the rural council has the power of decision over all land use rights within its territory with the exception of the following rights—mining, hunting, fishing and commercial forestry'. In the same decree, it is stated that:

...the rural council may pronounce on all regulatory measures whose implementation it deems useful and which are necessary in the land under its jurisdiction to ensure judicious exploitation of resources and effective protection of agrarian assets of all kinds.

Law No 80—14 of 3 June 1980 authorises the rural council to decide in all matters which the law has placed under its jurisdiction, especially with regard to combating fires and the practice of field burning; ways to control access to and use of water points of all types; the creation and establishment of livestock tracks within the rural community; and the development and use of all produce and firewood gathered from the wild.

Policy and Development Strategies

Rural development policy in Senegal is based on zonal stratification of production. This takes account of the potential of each ecological zone. These zones are then made the responsibility of specific regional rural development agencies (known as SRDRs) (see Map 11).

The development strategy for animal production is also based on the same stratification and its implementation is the responsibility of the relevant agency working in the region.

With the completion of the major dams, the Senegal river basin has been set aside for a variety of schemes, including closer integration of agriculture with animal husbandry and forestry, intensified cattle breeding and fattening, and the establishment of dairy units and poultry farms. The sylvo-pastoral zone, which contains almost a quarter of the nation's livestock and is traditionally given over to extensive animal husbandry, is the area set aside for intensification of cattle and sheep breeding.

The importance of agricultural by-products in the groundnut growing basin gives this region an advantage in terms of animal breeding and on-farm fattening. The strategy is based on a greater integration of agriculture with livestock

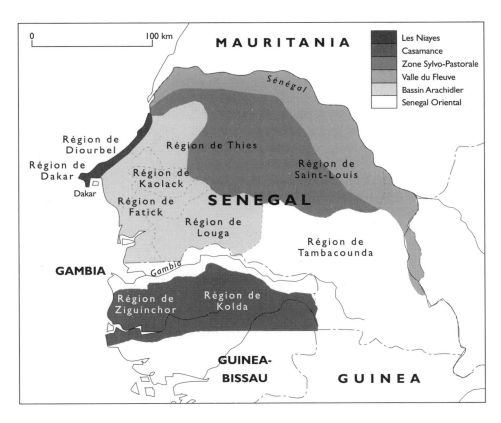

Map 11 Ecological and Administrative Regions of Senegal

through the use of animal traction, breeding of young cattle by GIEs (economic interest groups), and final on-farm fattening, as well as intensified breeding of small ruminants. In the southern and southeast regions, emphasis is placed on intensive breeding of tsetse-resistant *ndama* cattle. The Niayes Region is suitable for final fattening because of its proximity to markets and the large quantities of crop by-products available.

TRENDS IN PASTORAL DEVELOPMENT

The expressed aim of the state has always been to ensure better control of rural areas with a view to more effective use of natural resource potential. This has led to the adoption of a series of legislative and regulatory measures. The latter are characterised by attempts to persuade pastoral people to become sedentary and a desire to change their methods of herding and stock management.

In 1964, land reserves, which were a vital safeguard for the pastoral economy, subject to the vagaries of nature, were almost entirely devoted to the development of agriculture and support for intensive animal husbandry units. In this way 30,000ha of land were allocated for settlement to the Land Development and Exploitation Agency for the River Delta (SAED) in the middle of the Peuhl area in the Senegal river delta. The land surrounding the Guiers Lake was to be turned over to farming (under the responsibility of the Senegal Sugar Company and SAED) and the Senegalese Agricultural Development Agency (a cattle fattening farm which subsequently went bankrupt). This freshwater lake provided a watering place for local livestock which also used the floodplain pastures.

To the south of the sylvo-pastoral zone, an 80,000ha range was to be set up in the Doli Reserve. In order to relieve pressure on the groundnut-growing basin, the New Land Agency was also set up in eastern Senegal in the midst of pastures used by nomadic herders in the dry season. In the Dakar Region, the industrial horticulture agency, BUD Senegal, was established on customary lands of small farmers and Peuhl herders. The SOCA (Société Alimentaire), a private agro-industrial combine, was to take this over in 1989.

This policy of increasing agricultural production effectively authorised encroachment by farmers onto sylvo-pastoral reserves. For example, 10,550ha of forests were declassified in Deali and Boulel by Decree No 66—45 in 1966, and 45,000ha in Khelcom in 1991.

In order to conserve certain fragile ecosystems, the colonial authorities took steps to classify forests in the 1930s. The sylvo-pastoral reserves thus set up were areas of pasture where only food crops grown by the resident population were authorised. These conservation measures were reinforced by the forestry code (Decree No 65—078 in 1965 and Law No 74—46 in 1974) which regulated the use of rangelands in the classified areas. Legislative and regulatory measures were also taken to set up game reserves and national parks, of which the first, Niokolo Koba, was established in 1954. Unlike the sylvo-pastoral reserves, livestock was barred from the areas set aside to protect game, which currently cover about 1,020,000ha.

Donors Involved in the Development of Pastoral Regions

The main donors involved in the development of animal husbandry are listed in Table 5.1. Amounting to 2.7 billion FCFA (Francs Communauté Financière Africaine), the three year investment programme for this sub-sector represents about 1.8 per cent of the three year public investment programme (PTIP) for the primary sector and 1 per cent of the overall PTIP.

Table 5.1: Funding of Development Projects in Senegal, 1988 —1991

Project title	Donors	Cost (in million FCFA)
SODESP	CCCE/FED/FAC/BNE	1,821
PRODELOV	FAC/BNE	350
Revival of bee keeping	BNE	90
Establishment of fodder reserves	BNE	100
Buffalo project	USAID/BNE	246
PDESO	AID/CCCE/BADEA	89

CCCE: Central Economic Cooperation Fund
FED: Fonds Européen pour le Développement
FAC: Aid and Cooperation Fund
BNE: National Equipment Fund
PRODELOV: Projet de Développement de l'Elévage Ovin
USAID: United States Agency for International Development
AID: Agency of International Development
BADEA: Arab Development bank for Africa

This low rate of investment might look as if the state has abandoned its objectives for the livestock sector (an increase in meat consumption of 33 per cent plus the integration of animal husbandry with agriculture), but in fact it reflects a new strategy adopted for the development of the sub-sector. This new strategy calls for a disengagement of the state from production in favour of private operators. The state confines itself to:

- promoting animal healthcare to guard against major epidemics;
- fostering the training and organisation of herders;
- undertaking research and development, leading to an intensification of production and thoroughly restructuring the marketing system.

Regional rural development agencies are charged with supporting all the farmers and pastoralists in their catchment area. Animal husbandry, however, is seen as a secondary aspect of agricultural activities, except in specifically designated pastoral regions.

The pioneer lands, which the public authorities ceded to the development agencies, were mainly used for the production of rice, cotton, groundnuts and maize. The location of agricultural and irrigation schemes has often prevented herders from using the pastoral resources to which they previously had access—for example, the waters of the river Senegal and the floodplain pastures of Guiers Lake. The retreat of herders back to exclusively pastoral zones, to which access is frequently difficult, also means that in most cases they are unable to benefit from agricultural and agro-industrial by-products available in farming areas.

In the sylvo-pastoral zone, responsibility for extension work and development assistance to herders, as well as for the management and exploitation of the Doli Reserve as a state ranch, was given to the Livestock Development Agency for the Sylvo-pastoral Area (SODESP). Its strategy for the area was to be based on stratification of production at three levels: breeding in the Ferlo Region, rearing in Doli, and final fattening in Dakar. In taking on this role, SODESP agreed to address some of the state's major concerns with regard to herders.

The mobility of pastoralists, as well as their perceived minimal contribution to the national economy, has always represented a problem in the eyes of the public authorities. The herder no longer has his former freedom of movement, as he must now inform SODESP in advance so that the registered animals can be monitored. With this in mind, and in order to ensure greater control over grazing land in its catchment area, SODESP has taken steps to make herders pay for the water consumed by livestock at the boreholes which it operates.

When GIEs, which are responsible for managing the boreholes, were set up, the herders were extremely reluctant to cooperate, but the support of the public authorities has enabled these groups to be maintained—although their legitimacy still requires strengthening. A literacy programme was also launched by the agency to allow the GIEs to adopt sound management practices and protect the interests of their members.

In 1981, the Senegalo-German Agro-sylvo-pastoral Project in Northern Senegal (PSAEAS) was established around the Widou Tiengoli borehole. This programme was designed to mark out and develop 200-ha plots over an area of 1400ha. Stocking rates for each scheme were limited: 10ha per UBT (Unité de Bétail Tropical/Tropical Livestock Unit) in three of the plots and 14ha per UBT in three others. Each unit represents a live animal of 250kg in weight.

The success of this first operation enabled PSAEAS to proceed with an initial extension of 14,000ha in 1986. A second extension took place in 1989 with a scheme of 4200ha. Although the PSAEAS development model was accepted by the herders, they modified it to bring it into line with their own production objectives (Toure 1991). This meant that neither maximum stocking rates, nor destocking and the marketing of young weaned animals, the most important stipulations of the model, were adhered to.

The main beneficiaries of this project have been the most influential families. Inasmuch as 'plot holders' enjoy exclusive rights within the scheme and also take advantage of non-project areas on the same basis as those outside the schemes, feelings of frustration are growing. While plots within the scheme maintain low stocking rates, neighbouring non-project pasture areas are under heavy and rising grazing pressure.

In eastern Senegal, Decree No 76—1242 of 31 December 1976 entrusted the Textile Fibre Development Agency, under the aegis of the Livestock Development Project in Eastern Senegal (PDESO), with the management of about $13,000km^2$ of state-administered property in the area to the north of the Dakar—Kidira railway in the Tambacounda and Bakel departments. This decree granted agricultural areas and rangeland to cooperatives or groups of herders who must make use of them according to the rules laid down by the agency. The area has been split into pastoral units, each of which is provided with a 'management plan' and an elected 'management committee', responsible, among other things, for combating bush fires, managing grazing land, and choosing sites for the establishment of infrastructure (Dieme 1986).

The implementation of these management plans allowed land use to be regulated. Plateaux areas were set aside for rainy season grazing while the animals were allowed to come to the designated agricultural areas in the valley during the dry season. Pasture deemed to be in a degraded state was put out of bounds.

The participation of herders in the implementation of the management plans made them feel more responsible for their own environment. The committees to combat bush fires have become more dynamic, and discipline within and between pastoral units has been greatly improved.

Women

Although men are responsible for herding and managing the herd, some animals are the property of women. Furthermore, women are responsible for milking the cows, dairy produce, and the camp water supply.

Within the framework of the activities of PSAEAS in Widou Tiengoli, women were not, as a result of their social status, allocated pastoral plots. The project's activity did, however, affect their living and working conditions (Toure 1991). Almost all the households established within these schemes also have camps outside the scheme where some family members reside. In most cases the husband lives with his first wife on the plot while the second wife looks after the herd outside. The second wife is often obliged to take care of the children of the first wife as well as elderly family members.

The selection of animals to be brought into the project area is based on a concern to include animals belonging to both wives as well as the husband's livestock. On some farms, the women contribute towards the payment of dues. However, as a general rule, the initiative to destock young cattle is taken at the husband's discretion and the women are rarely consulted, even when this operation relates to animals belonging to them.

Apart from the disadvantages resulting from the break up of family groups and the assignment of some women to camps located outside the schemes, the overall effects of the project's work have been positive for women. For example, the time spent fetching milk has been reduced and milk production has increased.

The establishment of the SODESP feed supplement programme has improved the health and production of the herds in its care (shorter intervals between calving, increased milk production and so on). However, the fact that calves are taken away after eight months, or a year at most, represents a loss of potential income for the women (Pouillon 1984).

PRINCIPAL CAUSES OF ALIENATION AND CONFLICT

For pastoralists, livestock represent capital, including the herd as a means of production, using raw materials, fodder and water, and requiring the 'work' of one or more herders. This capital investment produces milk but also more livestock. Part of this production is reinvested. Calves born each year can be viewed as reinvested production; if they are left with the herd, they allow production to be increased.

Furthermore, livestock wealth provides the Peuhl pastoralist with economic power which gives him social standing and 'political' sway (Toure 1983). Indeed, the *djaarga* (a man rich in livestock) has a key role in Peuhl society. Large herds are also seen as a means of overcoming the inevitable losses caused by low rainfall and inadequate forage supply. Therefore land as such has no value. It is not a means of production, as it is for farmers. Its value lies in the vegetation and water that it 'produces'. These are renewable natural resources, but are not permanently located in a given place, so the best pasture and water must be sought each year and each season.

The Peuhl herders also prefer to produce as much as possible within the constraints of the available workforce. In the Sahelian environment, transhumant animal husbandry, as practised by the Peuhl, seems to be the most profitable economic activity with the lowest possible labour demand. The perception that pastoral people have of their own activities has promoted the view among the authorities that 'Peuhl nomads are elusive, arrogant, undisciplined and more attached to their livestock than to the frontiers of their new countries and are very mediocre citizens' (Bugnicourt 1977), or that they are 'opposed to change and aim simply to acquire as much livestock as possible' (Baker 1979).

Such assessments have very often meant that pastoralists have lost out to farmers in the struggle to preserve their natural rangelands from agriculture. The transfer of land ownership has in recent years favoured rich, influential city dwellers to the detriment of transhumant herders who become dispossessed of their rights to use land. This is the principal source of conflict and impoverishment for pastoral people.

Animal husbandry is also not considered to be a sufficiently productive form of land use by the government, with the result that some of the best Sahelian grazing has been destroyed in favour of cropping. And the fact that pastoral groups have not been given legal status also hinders the emergence of herder organisations which could guarantee their own economic and social development.

In the absence of definitive regulations for transhumant pastoral land use and land tenure, herders feel unable to undertake pastoral improvements, because their enjoyment of these is not guaranteed by security of tenure.

Access to Pastoral Resources

Water

Until recently, water points were run by the public authorities. With the increase in livestock numbers and the poor operation of most water points, where much of the equipment is old and beyond repair, the problem of water supply in the dry season has become more acute.

The first experiments in seeking contributions towards the running costs of boreholes began at those boreholes run by SODESP. The results have generally been positive, and as the state disengages from most sectors, it has realised the need to give local people greater responsibility in managing the infrastructure crucial to their production strategy.

A ministerial circular dated 1 January 1984 authorised the creation of committees responsible for managing pastoral boreholes. These are non-profit making bodies whose members (a maximum of 12) are nominated by the people in the borehole catchment area. The main objective is to minimise delays and deficiencies in the supply of fuel and lubricants necessary for the smooth running of the water pump. The members can also take responsibility for the maintenance of the pump and motor. The committee operates on the basis of dues paid by users and grants from public or private sources.

At boreholes managed by SODESP, the herders within the scheme have free access to the borehole and pay a watering tax on the same terms as credit for an input supply. Those outside the scheme, but residing near the borehole or outside the borehole catchment area, have to pay a watering tax assessed on the basis of the weekly requirements of each animal within the herd (Guèye 1985).

Pasture

Grazing areas are defined by a law on state-administered property, complemented by a law on administrative, territorial and local reform and the decree governing the rangelands. Decree No 80—268 of 10 March 1980 defines the conditions for the classification and declassification of natural pasture. Changes in classification can only be done after a detailed study including:

- a detailed map showing the location of villages, lands destined for agriculture, land destined for grazing, fallow or arable areas, forestry reserves, lands whose classification or declassification is requested, and the size of the village populations and herds of livestock;
- justification for the classification or declassification;
- minutes of the departmental commission meeting, etc.

In the pastoral herding system, grass is the exclusive property of those who use it even when it becomes scarce. It is not uncommon to see a herder leave his permanent camp to go elsewhere because transhumant pastoralists have come to settle near by.

In classified areas (forests and sylvo-pastoral reserves) the use of grazing land is regulated by the forestry code, Decree No 65—078 of 10 February 1965. Article D24 states that the passage of domestic animals in the reserves is authorised except in reafforestation or restoration schemes, in artificially replanted areas where the presence of animals could damage the plantations, and in naturally regenerating areas of forest. The felling of trees, whether or not of protected species, to feed livestock is forbidden.

In specifically designated pastoral regions, trimming or cutting of branches of fodder trees may be authorised under certain conditions as a user right.

Forestry Resources

Grosmaire (1957) reported that in the sylvo-pastoral zone 'sovereignty over wooded areas could be granted or allocated to certain communities by the civil authorities'. The Laobe, who used certain types of wood to manufacture tools and domestic implements, took advantage of such privileges. Similarly, a Moorish fraction, by virtue of political influence at the time, was allowed to harvest gum in a particular sector.

The Forestry Code of 1965 alters rights of access to forest produce. It stipulates in Article D19 that adjacent rural communities, or those who traditionally used the areas, are authorised to enjoy unrestricted user rights to the collection of dead wood, harvest of wild fruit, food or medicinal plants, gums and resins, straw, honey and any other user right acknowledged by classification decrees or orders.

Article D20 of the code additionally provides that the exercise of user rights is strictly limited to personal and family needs and not for commercial purposes. In return, these users are obliged by Article D21 to contribute on a pro-rata basis to the maintenance of the forest in which they enjoy user rights.

Livestock Capital

In the sylvo-pastoral zone, the herd is managed collectively but is the individual property either of members of the camp (*galle*) or of people from outside, such as friends or relatives (Santoir 1982). The family herd is a mixture of four elements: cattle owned by the head of the *galle*; cattle given to the wife (by her husband) on marriage; cattle belonging to children; and cattle belonging to married relatives living in the *galle*. All the animals are placed under the authority of the *galle* head who takes all important decisions on herd management. However, he can only dispose freely of his own animals and those ceded to his young children. He is responsible for the maintenance only of any other animals entrusted to him.

Santoir points out that small livestock are subject to more or less the same rules, but they do not have the same social and economic value as cattle, and they are frequently outside the control of the *galle* head.

In the livestock market, prices are fixed by the law of supply and demand. They do, however, vary according to season and are higher in the rainy than in the dry season.

ENVIRONMENTAL IMPLICATIONS

Since 1968, pastoral ecosystems in Senegal have, in common with most parts of the Sahelian region, suffered a serious reduction in rainfall. This has resulted in very high livestock mortality, and highlighted the degraded state of the natural resources within such ecosystems.

Natural plant formations have changed substantially. Bare surfaces have developed and evidence of wind and water erosion is increasingly visible. In the sylvo-pastoral zone, for instance, perennial grasses (for example, *Andropogon gayanus kunth*, a species of grass) have given way to other species which are less productive and/or less appetising to livestock. Primary herbaceous productivity has also fallen. Biomass production, that is, quantity of grass produced by surface area, used to exceed 3000kg of dry matter per hectare in certain places. It now sometimes fails to reach 500kg of dry matter per hectare.

In some areas, the shrubby steppe (coverage of woody species between 7 and 15 per cent in 1954) had become grassy steppe by 1980 (coverage of woody species below 2 per cent) (Barral, Benefice, Boudet et al 1983). Certain woody species, such as *Sclerocarya birrea*, *Commiphora africana* and *Grewia bicolor*, which are sought after by animals and used extensively by people, have become increasingly rare; on the other hand, other less desirable species, such as *Boscia senegalensis* and *Calotropis procera*, have flourished.

The pools which still provide drinking water for people and livestock in the rainy season are no longer filling up as before, while their capacity has often been reduced as a result of silting-up. There are several reasons why the pools do not fill up. Drought, livestock use and wind erosion are the chief reasons. Drought affects the volume of water, the presence of livestock causes the soil to pile up, and wind erosion creates sandy deposits.

Most of the wild animals (for example, lions, elephants, buffalo) have disappeared while others have become increasingly rare as a result of the destruction of their habitat and the reduced availability of water and food.

As a result of hydraulic, sylvo-pastoral, and veterinary schemes, the numbers of domestic animals (cattle, sheep, goats, horses, donkeys and camels) initially grew appreciably. They then fell sharply during the droughts of 1972 and 1984. Subsequent higher rainfall has allowed herds to be built up again rapidly.

Environmental changes are the result of the combined action of drought, overgrazing, and offtake for domestic purposes by pastoral people. These changes also depend on the nature and characteristics of different sites. For example, the reduction in density of woody species on sandy soils is largely attributable to drought, whereas on gravelly soil the principal factors are people and livestock.

The increase in bare soils appears to be mainly due to overgrazing and is found particularly where the landscape is characterised by the presence of depressions and temporary pools, at which livestock are watered.

Impact of Livestock

By grazing, livestock directly influence vegetation, the impact of which depends on the season. Grazing and browsing during the early growth cycle may hinder the regrowth of woody species and the flowering and seed production of grassy species. It is particularly around permanent settlements and water points used in the rainy season that grassy vegetation is affected. The first to disappear are rapidly germinating species because after the first rains livestock requirements are more than the available shoots can support. This means that slowly germinating species will form the greater part of vegetation which subsequently takes hold.

These direct effects are more noticeable in well-drained soils, as elsewhere the influence of run-off is a much more important factor on pasture composition: for the same type of soil with the same rainfall, vegetation may be entirely different as a result of changes in the availability of water.

Reductions in plant cover render the soils more vulnerable to the extreme force of the rains. This causes a crust to form on the surface of the soil and for the upper layers to become compacted. This results in impermeability and increased run-off, with a consequent decrease in the water content of the soil (Penning de Wries and Djitèye 1982). The crust which produces an increase in run-off, however, can also protect the soil against water and wind erosion. In Sahelian pastoral areas, the shape of the substratum is generally fairly even, so water erosion is minimal and, very often, the water forms only temporary pools.

The presence of animals on rangeland not only reduces plant cover but tramples the upper soil layers, causing subsequent deterioration. In sandy soils, degradation occurs mainly in the form of compacting the upper and pulverising the lower soil strata, the intensity of which diminish the greater the distance from the water points. In gravelly soils (shallow soils with poor drainage and more accentuated slopes), the combination of these factors with drought accentuates water erosion, particularly around water points.

In contrast to observations in other semi-arid regions, crusting of soils cannot, in sandy areas, be attributed to the effects of overgrazing and trampling (Barral, Benefice, Boudet et al 1983). It is actually in areas furthest from boreholes and even in parts where grazing is forbidden that the formation of thin surface crusts is most marked. This is largely attributable to drought and the lack of minerals in the soils.

Nitrogen and phosphorous are displaced from the pastures to areas where animals concentrate (camps, livestock compounds, around water points) and the tracks they use to reach these places. The consequence is a slow but inexorable depletion of vast areas already suffering from restricted vegetative growth. Wooded and shrubby vegetation is essential for daily life (firewood, buildings, fencing, domestic utensils) in the sylvo-pastoral area.

The Impact of People

As rainfall diminishes, the renewal of plant resources occurs ever more slowly. In theory, offtake of materials for domestic purposes should thus be reduced to ensure the sustainability of the most heavily used species. Unfortunately this does not happen in practice. Moreover, pastoralists are turning more and more to agricultural activities to make up the deficit in food supply (for example, to millet, sorghum, cowpeas and beref (*colocynthus citrullis*)). Fields are established by clearing trees and every year trees and shrubs are cut down to repair fencing.

In view of demographic pressures and the loss of fertility of former agricultural lands, families have had to search out 'virgin' or 'ownerless' land. This has meant encroachment on designated pastoral land, with serious consequences for the environment. In Senegal, this has reached alarming proportions, especially on the southeastern border of the sylvo-pastoral area in the groundnut-growing region.

Although the overall reduction in resources has meant that the Peuhl are no longer the excellent nature conservators mentioned by Grosmaire (1957), their settlement areas have few patches of denuded soil. By contrast, vast denuded areas are particularly noticeable around farmer villages where the size of the population means that fuelwood requirements are enormous. The same goes for areas cleared for agriculture, as it is often the sole source of income for these people.

In cultivated areas, water erosion has intensified as woody species have become rare. Tillage and harvesting of crops also encourages wind erosion during the dry season.

Factors Aggravating Ecological Tension in Pastoral Systems

The reduction in grazing areas, coupled with growing animal populations, is one of the major factors in the crisis which has afflicted pastoral systems. Agro-sylvo-pastoral developments between 1960 and 1990 (Table 5.2) show an increase in areas given over to cereals, as well as those set aside for national parks. This is equivalent to a reduction in rangeland for livestock whose numbers have more than doubled.

Table 5.2: Agro-sylvo-pastoral Developments in Senegal between 1960 and 1990 (.1000)

	1960/61	70/71	80/81	89/90
Cultivated area (hectares)*	1921	2284	2428	2145
National parks (hectares)**	260	0913	1020	1020
Livestock (UBT)+	-	2888	2628	3366

*   Agricultural service database
** Data from the National Parks Service
\+   Data from the Livestock Service

The essential problem is, therefore, access to pasture and the survival of larger herds condemned to division, dispersal and restricted mobility. The causes of reductions in grazing areas differ according to region (Ba 1982). In the Senegal valley, especially in the delta, the development of commercial farming is the main factor limiting access to rangeland. Off-season crops (rice, sugar cane and tomatoes) and the permanent developments which go with them have changed the way land is used. In the sylvo-pastoral area, the shrinking resource base is more likely to result from concentrations of livestock around boreholes and the frequency of bush fires.

In the groundnut-growing basin, the area given over to groundnuts has been constantly increasing. The expansion of intensive cropping has reduced fallow periods and increased the rate of clearance bordering the sylvo-pastoral zones (upper Sine valley, Kaffrine department and so on). The extension of halomorphic land in the Sine valley has halted the movement of herds towards the coastal regions. Furthermore, the reafforestation policy pursued over recent years means that plantations of eucalyptus and other species that are not palatable to animals (Thiès and Sine Saloum regions) are gradually replacing the wooded rangeland areas.

In eastern Senegal, the maintenance of large herds is becoming increasingly difficult due to the expansion of both cultivated (especially groundnuts and cotton) and reserved areas. The Niokolo-Koba national park, which covered

260,000ha in 1954, now stretches to over 930,000ha. Some villages have been obliged to move, often to regions that are ill-suited to herding.

In Casamance, bush fires, land clearance and hydro-agricultural schemes have restricted agro-pastoralism, the mainstay of this region. Areas given over to groundnuts almost doubled between 1950 and 1980. Commercial cotton plantations, introduced at the end of the 1960s, covered 12,600ha in 1989/90. Moreover, cotton delays the use of post-harvest pasture because of its longer cycle.

Finally, in the Dakar Region and the coastal areas, urbanisation, tourism and the extension of vegetable and fruit cropping (accompanied by the installation of fencing) have combined to push herds further into the hinterland.

CASE STUDIES

Land tenure issues constitute the greatest problem in the Senegal river valley. Much has been written on this problem. A summary is given here of the studies undertaken by Niane (1987, 1990).

The Senegal river valley has a high concentration of people and animals. Sedentary and so-called 'semi-nomadic' people are livestock owners. However, the latter look after most of the animals. Both pastoralists and sedentary people own land, as they are often from the same family.

The delta had a low population density before irrigation schemes were set up around 1963. It was an area of common grazing, and access to the Senegal river and its tributaries was easy in the dry season. However, the dams associated with the irrigation schemes have promoted a stampede towards the floodplains from outside the country and from all regions of Senegal by wealthy people and former land owners, even though water control and irrigation from, for instance, the Manantali dam, are not yet fully effective. The result has been numerous conflicts that are far from being resolved.

The alluvial soils of the valley have always been coveted and argued over, but especially since recent droughts have forced governments along the river (Mali, Mauritania and Senegal) to proceed with construction of the Diama and Manantali dams. Since the advent of irrigated cropping schemes (sugar cane, rice, tomatoes, etc) conflicts have intensified and have even resulted in violent quarrels between pastoralists and sedentary farmers, who accuse the animals of causing damage to dykes, sluices and irrigation channels. These conflicts are never resolved to the satisfaction of the Peuhl, who are described as 'crop destroyers'.

In most cases local elected representatives (who tend not to be landowners) are rarely concerned with the fate of the electorate, which is 90 per cent illiterate.

These representatives often have little belief in community property and often turn out to be more despotic than traditional landowners, some of whom have made political alliances with the current rulers to save their own land, while at the same time speculating in land belonging to weak or absentee owners.

Economic and religious forces also exert pressure on the political system, while local people are fiercely determined to do everything they can to conserve their land. Moreover it is this fierce opposition by local people which gave rise to the war between Mauritania and Senegal. In any event, conflicts persist and the implications are uncertain.

Pastoralists remain marginalised and their way of life threatened by large farms or ranches where local farmers and rich people from the north have joined forces. Pastoralism is doomed to die out unless urgent consideration is given to establishing communal farms or ranches where herders can learn about modern technology, backed up by literacy training for the young herders to take over from their illiterate and disoriented elders—the basic elements of the current land tenure crisis, leaving aside specific variations from one department to another, and from one ethnic group to another, are also determined by technological innovations and the socioeconomic and cultural readjustments they demand.

Another flash point in the struggle for access to land can be found in the groundnut-growing region dubbed the 'Khelcom forest affair'. Quite recently, a classified forest was declassified for the benefit of a religious Mouride leader for groundnut farms. This meant that a whole pastoral community was displaced and at the same time lost the use of pastures where they had been bringing their animals for more than 40 years.

The Mbecque or Khelcom forest is located about 300km to the east of Dakar. It covers an area of 75,000ha and is bordered to the north by the Doli ranch, to the east by the Sine Saloum sylvo-pastoral reserve, and to the south by Kaffrine Department. The forest contains 35 ponds which used to retain rainwater for more than two months after the rains. Deer, partridges, guineafowl, hyena and jackals are among the wildlife found there. Pastoralists have been settled in the forest since 1950 in 370 villages/compounds holding more than 6000 people and 75,000 head of livestock. This forest was also used by transhumant animals elsewhere in Senegal.

Under the new legislation, 45,000ha (or 60 per cent of the forest area) have been withdrawn from traditional grazing by herders and are being converted into farmland. The 35 ponds are all located in the area covered by the legislation. The most urgent problem for these pastoralists is to find somewhere else to go.

How this unexpected dispersal will affect the rather strained relationship between pastoral and farming communities in the medium term presents another problem. How are those pastoralists residing in the non-declassified area going to live alongside the community which now controls their former grazing?

Traditionally, use of land in the sylvo-pastoral area of Senegal was governed by the rule of *ouroum* (Grosmaire 1957, Barral 1982, Ba 1982 and Niang 1982). The Peuhl distinguished owned land (*diei*) from land which was not owned (*ladde*). *Diei* is land located within the catchment area of a pond or system of ponds. All land belonging to a single owner was part of the territorial unit known as *ouroum* and the people who lived there knew exactly where the borders were. Within each *ouroum*, a kinship group lived in camps and anyone outside the community who wished to settle within the *ouroum* had to seek the authorisation of the land chief. In those days the sylvo-pastoral area was only occupied during the period when the ponds contained water. The herders subsequently moved their animals towards the river valley and neighbouring pastures.

These movements to and from the river valley began to tail off when the pastoral water supply programme began in the 1950s in the Ferlo region, encouraging more permanent settlement by herders (Barral 1982). The activities of the livestock service (combating disease and predators) and the forestry department (combating bush fires) have increased livestock numbers. The lifting of constraints on animal husbandry went hand in hand with a weakening of solidarity among, and a decline in the authority of, traditional chiefs. As respect for the *ouroum* rule increasingly diminished more and more, so the law on state-administered property (in 1964) and that concerning administrative, territorial and local reform (in 1972) took over.

Pastoral regions play a part in the national economy through their animal products (meat, milk, hides, etc), forest products (wood, gum, etc) and game. Wincke (1990) estimated that in the sylvo-pastoral zone (39,000km^2), this represents 64 billion FCFA, producing annual interest of 20 billion FCFA on pastoral activities, or 29 billion FCFA on agro-pastoral activities.

Relating these figures to all of Senegal (192,000km^2), and considering only animal production on the basis of traditional animal husbandry, Wincke states that poor management of natural resources represents a gradual loss for the country of 323 billion FCFA, plus 100 billion FCFA annual interest. This shows that the preservation of pastoral ecosystems and indeed an increase in their productivity is in the interests of both herders and the whole country.

The predominance of pastoral activities in Sahelian ecosystems is due mainly to the fact that environmental conditions (for example, limited water resources and low fertility soils) are less supportive of other activities. From an economic point of view, animal husbandry practised by pastoral people seems to be the most profitable way of using natural resources and, assuming some limitation on the number of animals, the least threatening to the environment.

In areas with considerable production potential as a result of irrigation schemes (river valley) or high rainfall (in the south), the problem should be seen in terms of the need to integrate agriculture and animal husbandry. Separation of land use systems in itself would be valid, but problems are likely to arise in implementation. The exclusion of other activities such as livestock keeping is not conducive to optimising production or improving the food security of the population.

PROCEDURES FOR RESOLUTION OF CONFLICTS

Conflicts over land are generally due to settlement of farmers on traditional rangelands or the violation of cultivated land by pastoralists. These conflicts have almost always gone the way of the sedentary farmers, who were even helped at times by the public authorities. Ba (1982) concluded that the Mouride brotherhood was used by:

> ...*the colonial system to achieve a constant expansion of the groundnut-growing area. No stone was left unturned: deliberate administrative laxity, hundreds of hectares of land ceded, grants, assistance, extension services, handing over mechanical equipment to religious leaders, declassification of classified forests, massive seed distributions etc.*

Seeing the monopolisation by agri-business of land belonging to indigenous people around Lake Guiers, Niang (1982) noted that:

> *The most striking thing was that despite large scale land transfers, the public authorities were slow to react. One can only conclude in such cases that the course of events is not yet going against the State's objectives in respect of agrarian policy; indeed one would be tempted to think that the State's failure to intervene to restore order in the region is due to the fact that everything is going to according to a scenario planned and favoured by the State itself.*

This lack of reaction from the administrative authorities was also reported by Santoir (1983) in the groundnut-growing basin, but this is more likely to have been due to the fact that the administration did not have any means of control. In the event of conflicts, the legislative provisions of Decree No 80—268 of 1980 dealing with rangeland organisation are supposed to be applied. This decree authorises commissions set up for each administrative unit (region, department, *arrondissement* and rural community) to adjudicate in disputes over land tenure. Each commission includes representatives of the administrative authority, technical services, and farmer and herder cooperatives. These commissions are responsible for:

- handling papers relating to classification or declassification of land;
- assisting the rural council with marking out areas for pasture, rangeland, livestock routes and agricultural, borehole and pastoral developments;
- reconciling herders or livestock owners and farmers. Should the conciliation fail, common law tribunals are responsible for settling disputes.

In practice, however, land disputes are often settled at a local level by 'elders', notables, or local elected representatives. In PDESO's area of operation, agreements have been worked out with herders in neighbouring areas over terms for mutual access to their pastoral units.

Representation

It is certainly not an easy matter for a pastoral people, most of whom are illiterate, to comprehend the many laws, decrees, statutes or their enforcement orders. Thus, at Widou Tiengoly in the sylvo-pastoral zone, Toure (1991) observed that:

> *The allocation of plots within the PSAR scheme is seen by the people as implying recognition of ownership rights. Such rights granted to a non-local pastoralist would (therefore) as far as local herders are concerned, constitute a serious threat to their interests.*

The author goes on to say:

> *Claims made by the people...do not take account of the legal provisions governing the area. Widou Tiengoly is part of the area classified as a sylvo-pastoral reserve for whose management the forestry service is responsible. Particular user rights apply to these areas where the rural councils do not have the jurisdiction granted to them in country areas.*

This example makes clear the extent to which pastoral societies, in common with the whole of rural society and even some administrators, are unaware of most legislative and regulatory provisions dealing with land tenure. The disadvantaged position of pastoral people in relation to the settling of conflicts is compounded by their inadequate representation within decision-making bodies.

As early as the colonial era, the Peuhl found that their lack of sociopolitical cohesion left them subject to domination from colonial and certain sedentary groups. Political officials selected by the colonial administration to become heads of the canton or province came almost exclusively from among the sedentary population (Ba 1982).

In the rural community of Mbane, Niane (1984) stated that the 21 rural councillors included 11 Peuhl (that is, an absolute majority), nine Wolof, and one Moor. Despite this advantage, the Peuhl did not even succeed in retaining the rangelands that give access to Lake Guiers. They were pushed out by the southward expansion of the fields which they themselves, 'exercising full sovereignty', allocated to the various agencies which put in requests (Senegal Sugar Company, Senegalese Agricultural Development Agency). A distinction has to be made, therefore, between legislation and the actual jurisdiction of the rural council and the president of the rural community. This case, just one example among many, shows to what extent the rural council's power is more nominal than real.

Herder Organisations

In view of the perpetual conflicts in which they become involved, especially with farmers, pastoralists have been trying since 1947 to set up representational organisations which will give them a voice. The demands of the so-called 'Peuhl union' at its congress in 1957 concerned:

- the struggle against the brutal invasion of rangeland by farmers;
- the incrimination of the Economic and Social Development Fund, held responsible for replacing animal husbandry with agriculture;
- the denouncement of farmers thought to be planning to take over the boreholes.

The plan of action of the 'Herders Mutual Aid Society', set up in Senegal in 1965, contained the following objectives:

- organisation of herders' activities by setting up a pastoral zone which would free them from the need to watch their animals constantly, a task usually undertaken by youngsters who were thus deprived of schooling and condemned to illiteracy;
- sedentarisation of herders in order to dispel the mistrust of farmers.

Herders imitated other professional groups in rural society and set up cooperatives. These herder cooperatives were grouped into departmental, regional and national unions. They were intended to enable herders to gain access to bank credit on the same terms as agricultural, forestry, fishing and other cooperatives.

The revival programme for the 'Union of Herder Cooperatives of Dagana' provides for:

- participation in decision-making about location of livestock camps and the marking-out of grazing, agricultural and watering areas;
- an increase in the number of boreholes in the sylvo-pastoral area.

The Union of Senegalese Farmers, Herders, and Horticulturists, set up in 1977, was designed to protect the professional interests of its members and to bring about good relations between the various rural protagonists. It encountered opposition from the administrative authorities as well as the Union of Peasants, Pastoralists, and Fishermen (known as the three Ps).

None of these organisations, set up with or without the help of the public authorities, was really able to act effectively to further the aspirations of pastoral people. They were frequently taken over for political ends and now barely function. Under the new agricultural policy, the state has established structures that might be able to take on such a role. Under the aegis of Law No 84—37 of 1984, herders were to come together in GIEs, constituted into national, regional and departmental federations. In 1989, more than 1000 GIEs were set up, but only about 100 are functional and about 900 million FCFA have been granted to them under a credit scheme. This is the first time that herders have had access to credit. However, compared with farmers and fishermen, the amount of such credit is extremely low. How can the pastoralists use this credit to 'develop' or make land more productive in order to be able to claim rights over plots of land?

RECOMMENDATIONS AND CONCLUSIONS

Proposed Solutions to Conflicts Relating to Pastoral Land Use

Strict Respect of Regulations Governing Land Tenure

Even though the legislative and regulatory provisions currently in force do not enable land to be taken into communal or individual ownership for pastoral purposes, they do define quite clearly the norms governing the use of grazing areas. It is therefore essential that all these provisions should be respected, to avoid conflicts as well as to deal with disputes that do arise between pastoralists and other producers. Any change in the status of an area of land (classification or declassification) could therefore be considered by commissions on which pastoralists are properly represented.

The rural councillors must also be better informed about their role. They need genuinely to represent the herders.

Establishment of a Multi-purpose Land Use Policy

Areas devoted to extensive herding have been greatly reduced while numbers of livestock have increased. A strategy for improved multi-purpose land use therefore needs to be established.

Access to National Parks

In this context, the example of Kenya, where giraffes share the land with domestic herds (Bille 1991) shows that the needs of wildlife and livestock are not incompatible. But to what extent the trend towards excluding livestock from game reserves could be reversed needs to be identified. Discussions on this point are said to be under way within the National Parks Service.

Encouraging the Use of Post-harvest Pasture and Fallow Land

Arrangements encouraging maximum use of agricultural residues by herders should be made. Farmers should allow herders to come into their fields after the cropping season instead of quickly burning off harvest residues; equally, herders should respect the schemes set up by the farmers.

Proposed Solutions to Rangeland Degradation

The notion of 'making land productive', as presently defined, excludes the use of land for pastoral purposes. Consequently, pastoralists have no interest in investing in the conservation of the communal environment. The establishment of pastoral units should therefore be extended. As has been done in eastern Senegal, each herder community would thus be responsible for the use of all pastoral resources in a given territory. This territory would be allocated in accordance with the size of the human and animal population and its natural resources. Any decision concerning use by people from outside should be made by the pastoral unit.

Proposed Research Concerning Land Tenure Issues

History has shown that pastoralists are almost always the losers in the struggle for access to land. The objective of the political authorities—food self-sufficiency—invariably means crops and does not seem to be open to negotiation. The rangeland is thus likely to continue to shrink.

Faced with this situation, pastoralists will be able to continue their activities only if they adapt to the new environmental conditions. An example can be taken from Kenya, where Maasai pastoralists are:

> ...both traditional and modern: they send their children to school as often as possible, keep abreast of progress in veterinary care, manage large scale water resources and directly market a major part of their production. They were the first to take part in cooperative herder groups and even to invest in 'group ranches'. They practise communal ownership of the rangeland they use but retain individual livestock management. They are very adaptable and indulge in numerous secondary activities. (Bille 1991)

Research should therefore be undertaken to enable pastoralists to take better advantage of the present situation, such as how to maximise fodder potential in non-pastoral areas, and the pre-conditions for establishing communal grazing areas. A better knowledge of pastoral water point management and herder access to credit should also help to improve pastoral production.

The following research topics are suggested:

1. How to manage multi-purpose land use: for each administrative or ecological region, livestock numbers must be contrasted with the availability of pasture fodder, post-harvest fodder, fallow land and forestry and wildlife reserves. From this, proposals can be made to determine the potential for permanent or temporary use of the various types of grazing and the appropriate measures taken.
2. The pre-conditions for establishing communal grazing areas—this will mean assessing the current position of pastoral land in each administrative or ecological region, establishing the typology of these pastoral lands according to the traditional and current users and assessing the various strategies for managing rangeland, as well as considering the mechanisms for defining and promoting the terms for herder participation.
3. The involvement of herders in pastoral water point management—animals often water and graze outside their immediate 'home' area. When ponds fill up in the rainy season they are taken to drink there and they quite often go to graze in other borehole areas in the dry season. It would be interesting to examine how borehole management committees have evolved and the part played by different kinds of herd owner.
4. The effect of access to credit on the development of pastoral production and the effect of credit already granted on the development of pastoral production—in the light of the results, steps should be taken to strengthen and/or redirect the credit strategies.

[1] The Pastoral Network (REPA) is an organisation which brings together research and development officers from national and international agencies. It seeks to provide a framework for discussion of the problems facing pastoral people in Senegal.

6

SUDAN

ME Abu Sin
Department of Geography, University of Khartoum

Map 12 Sudan

INTRODUCTION

Pastoralism in Sudan (Map 12) involves about 20 per cent of the population (Table 6.1) and accounts for almost 80 per cent of livestock wealth. In the early 1980s there were nearly three million head of camel, over 20 million cattle, nearly 19 million sheep, and 14 million goats (Table 6.2). It is estimated that in the

Table 6.1: Nomadic Population by Regions, 1983 Census

Region	Number	Per cent of total population of each region
Northern	50,269	2.3
Eastern	558,676	25.5
Central	244,233	11.2
Kordofan	781,039	35.6
Darfur	469,744	21.4
Khartoum	88,000	4.0
Total	2,191,961	100.0

Source: Department of Statistics (1984) Regional Ministry of Finance and Economics, Eastern Region Population (1983) (in Arabic)

year 2000

Table 6.2: Livestock Numbers by Province, 1981/82

Region/province	Cattle	Sheep	Goats	Camels
Northern				
Northern	19,064	224,725	178,595	127,649
Nile	58,779	312,824	321,558	69,804
Eastern				
Kassala	864,388	1,824,882	1,130,836	671,295
Red Sea	49,062	257,077	543,995	112,952
Khartoum				
Khartoum	76,465	340,934	525,643	16,239
Central				
Bleu Nile	1,209,500	1,238,761	726,784	44,663
Gezira	677,540	1,396,414	1,398,202	172,627
White Nile	2,102,295	2,534,561	803,602	92,048
Total	5,050,793	8,140,188	5,628,715	1,307,177
Percentage	24.5	43.7	40.7	46.8

Kordofan				
N. Kordofan	1,260,015	2,836,379	2,058,281	1,006,542
S. Kordofan	1,972,948	952,952	850,906	2,125
Darfur				
N. Darfur	1,219,615	2,618,230	1,459,208	267,246
S. Darfur	3,677,827	1,414,442	1,365,817	167,937
Total	8,130,405	6,822,003	5,734,212	1,443,850
Percentage	39.3	36.6	41.5	51.8
Equatoria				
E. Equatoria	309	1,458	24,518	-
W. Equatoria	1,022,647	1,050,269	293,996	33,603
Bahr El Ghazal				
Bahr El Ghazal	1,650,712	824,583	738,519	-
Lakes	942,152	382,454	3,715,787	-
Upper Nile				
Jonglei	1,888,491	200,473	561,457	-
Upper Nile	1,920,143	1,202,554	459,502	5,817
Total	7,474,454	3,661,761	2,451,570	39,420
Percentage	36.6	19.7	17.8	1.4
Total	20,661,952	18,623,982	13,814,497	2,790,447
Percentage	37	33.3	24.7	5

Sources: Adapted from Obeid (1978), from South Matrur (1987) 'Planning for Sudan's Livestock Development: an analysis of the sub-sector towards 2000' Arab Authority for Agricultural Investment and Development (AAAID). Also appeared in Danida, Ministry of Foreign Affairs (1989) 'Environmental Profile: the Sudan' p26

camel and cattle numbers will be at roughly the same level, but that sheep numbers will increase to nearly 50 million, and goats to over 26 million (Table 6.3). These animals are almost entirely concentrated in ecologically marginal and semi-marginal areas under communal land tenure systems. These areas also contain zones of largescale agricultural development projects, traditional farming systems and areas of permanent settlement, based on different tenure systems.

Table 6.3: Livestock Population Estimates (in Millions), 1917–2000

Year	Cattle	Sheep	Goats	Camels
1917	0.753	1.266	1.216	0.234
1946	3.4	5.00	4.00	1.20
1962	9.10	8.66	6.85	2.00
1974	14.20	13.4	10.50	2.70
1983 (m)	21.38	29.27	13.39	2.80
2000 (m)	21.54	49.77	26.50	2.80

Sources: Adapted from Obeid (1978), from South Matrur (1987) 'Planning for Sudan's Livestock Development: an analysis of the sub-sector towards 2000' Arab Authority for Agricultural Investment and Development (AAAID). Also appeared in Danida, Ministry of Foreign Affairs (1989) 'Environmental Profile: the Sudan' p26

Note: annual growth rate 7.5 per cent in case of cattle; 3 per cent for sheep; 2.86 per cent for goats and 0 per cent for camels

These tend to be supported by the government and reflect land legislation that favours non-pastoral activities.

The interaction of climate, soils, topography and drainage creates a succession of different environments for which competition between pastoralists, farmers and government, as the largest investor in largescale agricultural projects, is fierce.

Rainfall is the main factor influencing the distribution of populations and livestock. It ranges annually from 75mm in the extreme north to 1500mm in the extreme south. There are five ecological zones with variable grazing potentials (see Map 13):

1. desert;
2. semi-desert;
3. low rainfall savanna and sanddune area, or *goz*, west of the Nile;
4. low rainfall savanna on clay in the east;
5. high rainfall savanna in the montane and in the flood region (see Maps 14—16).

Different livestock species and breeds tend to thrive in the different ecological zones that have distinct grazing qualities. In Sudan, the natural range is generally poor in nutrients and so it is not easy to find suitable locations to accommodate livestock all year round. Seasonal migration is therefore adopted to remedy shortages in pasture and water, to avoid biting flies and muddy conditions, and to escape mechanised or irrigated agricultural projects where pastoralists' admission is prohibited (Map 17).

Most pastoral lands in the Sudan are associated with a particular homeland, or *dar*, defined by customary rights. Within the *dar*, grazing is communal (Map 18). Competition for water and pasture since the condominium period of Anglo-Egyptian rule (1896—1956) led the independent government to confine tribes to their traditional *dar* or territories, suppress conflict, remedy hardship and encourage participation in economic development (Lebon 1965). This arrangement was disturbed by the 1970 Land Registration Act.

The northern arid and semi-arid zones favour camels over other species of livestock (Map 19). Their natural habitat is north of 14°N. After the drought of 1973, the camels migrated south to latitude 10°N even though this area is unsuitable for camels because of vector diseases (for example, trypanosomiasis from the tsetse fly).

The ownership of camels in these two zones is vested in the following groups:

- Eastern Sudan: Beja tribes; Rashaida and Ababda of the Red Sea; Shukriya; Batahin; Lahawin and Rashaida of Kassala.
- Western Sudan: Kababish; Kawahla; Hawawir; Shanabla; Hamar; Maganin; Meidob; Ziadia and Zaghawa of the northern areas, Kordofan and Darfur.

Camels in western Sudan have three grazing migrations: rainy season, winter and summer. Summer grazing areas comprise Khowi, Rahad, Nahud, Nyala, Daein, Wadi Howar and Wadi El Melik. During the rainy season they congregate in Hamrat el Sheikh, Syala, Um Inderaba, Mazroub, Um Badir, Malha, Kutum and Melit. Camels of the Red Sea areas move between the numerous *wadis* or *khor* (seasonally flooded water courses) as far west as the Atbara. Camels in eastern Sudan cover a much shorter migration between the rivers Atbara, Rahad and Dinder, and up and down the Red Sea hills.

During the past three decades frequent drought has resulted in the decline of pastures in the traditional *dars* and southern areas. Nowadays the nomadic camel-owning tribes of northern Kordofan penetrate earlier and far deeper into the Nuba Mountains and beyond, in search of pasture and water. This often brings them into conflict with rival tribes. This is also the case with the semi-nomadic Zaghawa of northern Darfur who seasonally move close to Bahr el Arab.

The Beja of eastern Sudan are pastoral nomads who move as little as possible

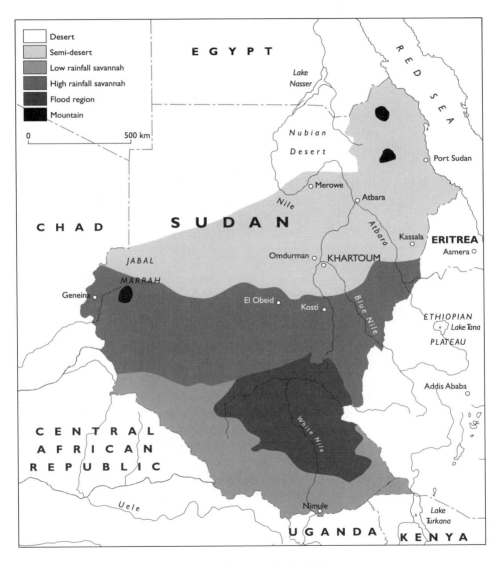

Map 13 Sudan Ecological Zones

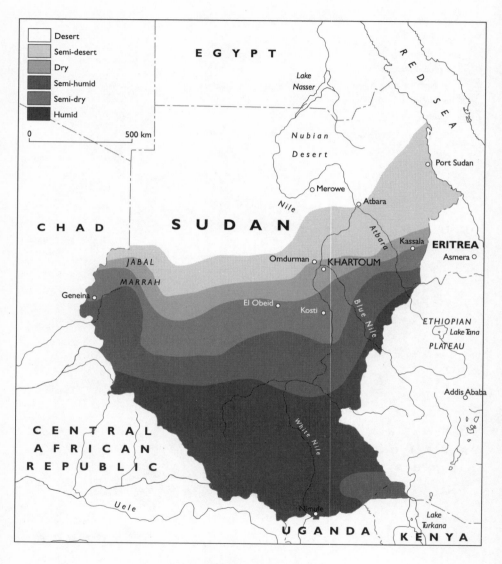

Map 14 Climatic Regions of Sudan

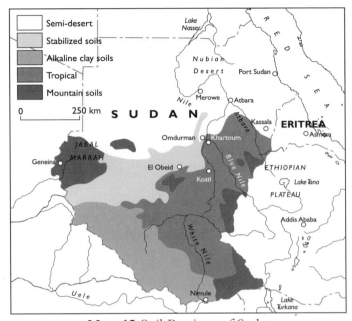

Map 15 Soil Regions of Sudan

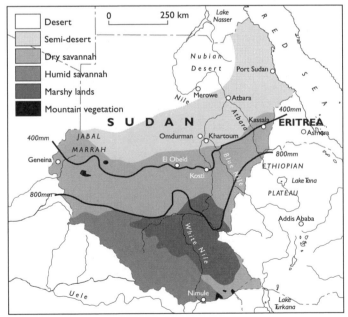

Map 16 Vegetation and Rainfall of Sudan

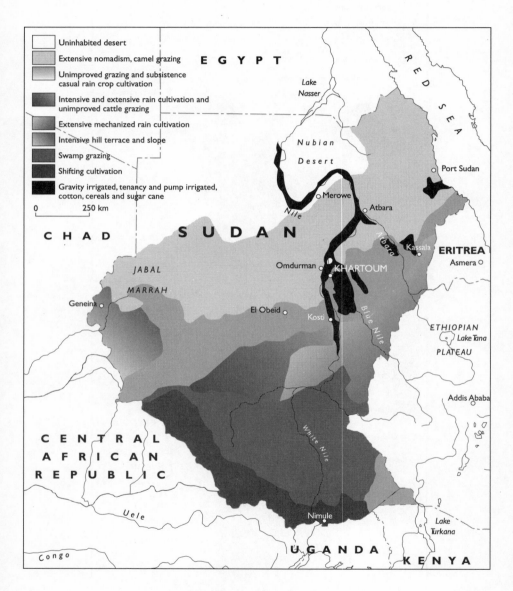

Map 17 Land Use in Sudan

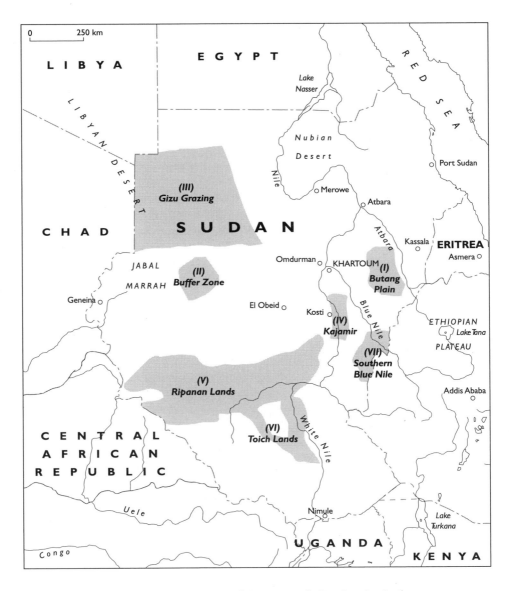

Map 18 Major Areas of Communal Grazing in Sudan

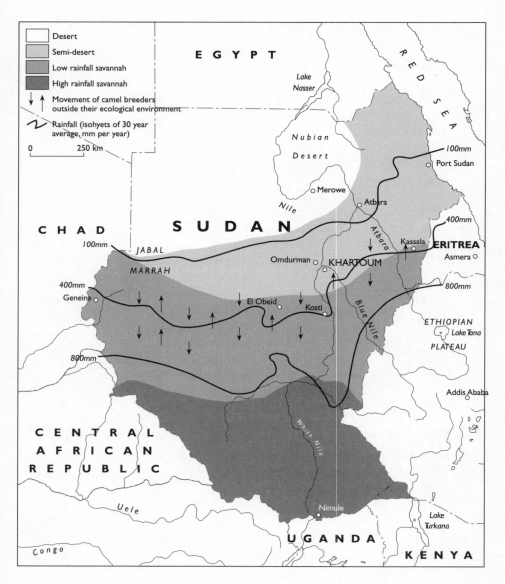

Map 19 The Camel Environment in Sudan

(from the Red Sea hills to the coastal plain in summer). Movement is restricted to a short radius from their wells. Only when drought persists for several years in succession do they move further, as they had to during the 1984—87 drought. The Beja move outwards from their traditional valley wells after the rainy season, looking for pasture on the hills, and migrate back to the water courses where they have permanent wells towards the end of the dry season. Sheep follow the same pattern as camels. They move to Gash and Tokar deltas where areas of flood irrigation offer crop residues.

In the northern drier parts of the country along the Atbara River live the Bisharin with their camels, sheep and goats. In the centre of the Red Sea hills live the Amarar who are semi-nomadic. South of this arid zone live the Hadendawa, along the Gash and Tokar deltas. They benefit from the profits of cotton cultivation which they invest in livestock. They move to uplands during winter and into valleys during summer. The Beni Amir live on the border between Eritrea and Sudan, which they frequently cross, especially during winter when rainfall is greater in Eritrea, and to the Red Sea hills and coastal areas in the rainy season.

The Shukriya tribe of the Butana (between the Blue Nile and Atbara River) own camels and sheep, as do the Rashaida and Lahawin. Batahin, Kawahala, and Rufa'a el Shariq are minor tribes in the same area. These tribes graze camels and sheep mainly on a grass with relatively high protein content, locally called *sheha*. During the dry season this clay soil plain dehydrates, which necessitates movement westward to water courses and wells near the hills. During the short rainy season the nomadic tribes in this zone disperse widely over the Butana and further west, as long as temporary water supplies exist. Wells in central Butana are used only by the Shukriya during the long dry season. Some of the Shukriya have to move near the Blue Nile and Atbara River, and irrigation schemes of New Halfa and Rahad for water and crop residues. Access is made easy for those who are tenants of these schemes.

Although many camel-owning tribes raise sheep in the arid and semi-arid vegetation zone, it is believed that half of the national sheep herd is located in the central belt (low savanna and high rainfall savanna). The Kababish have hardy sheep which can withstand their long migration. The Hamari and Kababish sheep can graze as far as southern Kordofan during the dry season, and the Qoz in the central part of northern Kordofan during the rainy season. In the dry season the watering interval is five to seven days in cool weather, and two to three days in summer. The grazing distance between water points is 20—30km, resulting in an overlap in the radius of water sources that leads to overgrazing.

Sheep from northern Darfur have started to migrate further south because of drought in the north. Traditionally they moved seasonally to the Gezu and in summer to their homelands in northern Darfur. Lack of water in northern Darfur is a more acute problem than in northern Kordofan. Where rainfall is low and vegetation is less dense, the incidence of tsetse fly decreases, and cattle and sheep may be reared safely.

Cattle, like sheep and camels, are kept at the *dar* most of the time within the central belt zone and the high rainfall savanna zone. In the central zone there are four groups of cattle-owning tribes:

1. those using the White Nile grazing areas during the dry season, for example, the Ahamda and Selim;
2. the Hawazma and Messeiriya Zurug living within the Nuba Mountains—often coming in contact with Nuba cultivators and mechanised agricultural schemes;
3. Messeiriya Humr, Rezeigat, and Habaiya of southwest Kordofan and southeast Darfur;
4. the southwest Darfur group, comprising the Fellata, Messeiriya, Beni Halba Ta'aisha and Beni Hussein.

The first group migrates southwards and towards the river during the dry season. During the wet season they move east and west to open pastures. In many cases they come into conflict with traditional peasants and settlers in this area. The main cattle-owning nomadic tribes, the Baggara, live further south below a band of cultivated land from Gedref to Wadi Azum. Living in *fereigs* (camps) on the plain between latitudes 10°N and 12°N, they often move northwards during rains and southwards during the dry season, covering long distances to reach scarce water and grazing. Heavy losses are experienced during droughts, and by the time herders recognise drought it is too late to save cattle (Bashari 1985).

South of latitude 10°N, however, the limiting factors for cattle grazing are the biting flies and muddy conditions. The pattern of migration is complex with short distances to cover. In this sub-humid region live the Nilotic pastoralists—the Dinka, Shilluk, Nuer, Mandari and Murle. A decade ago these tribes used to own over 30 per cent of the country's cattle. Now, with the political turmoil in the south, more than half of that wealth has probably been lost. These tribes have an annual cycle of migration (Lebon 1965). Their short distance migration from the unflooded highlands of the southern clay plain to the *toichs* (floodplains that act as dry season refuges) during the dry season is governed by drainage conditions, prevalence of biting flies and availability of pasture. At the first rains herds move from river floodplains to the highlands. The reverse occurs during the dry season. The Dinka have relatively longer distances to cover to the areas of Bahr el Arab and Bahr el Ghazal. This often brings them into contact with the

Baggara of southern Kordofan and Darfur and results in conflicts over grazing land. The latter tribes share dry season grazing with the Dinka and Nuer.

Goats are distributed all over the Sudan. Nomadic tribes, especially camel owners, have few goats, because they are unsuited for long distance travel.

Before the droughts of the 1970s and 1980s all pastoral livestock, in all the areas, moved northwards during the rainy season. During the dry season the livestock returned south to the permanent water sources in the same zone or even further south into the savanna zone. Camels and sheep normally grazed in the northern half of the country, cattle in the southern half. In western and eastern Sudan, movement is generally north to south; in the southern regions migration is between swamp and highlands within the *dar* in the dry season, and northward in the wet season.

Since migration is dictated by ecological and environmental factors, drought always results in longer migrations southwards, especially for camels in the dry season. Though cattle have restricted movement in the central belt and the south, in Kassala and Gedaref states they have a longer migration than in the northwest due to the presence of agricultural schemes which have cut across the traditional migration routes. Additionally, the expansion of cultivation into grazing areas is rapidly degrading resources. Some Beja have dairy farms around Kassala, Gedaref and Port Sudan.

Sheep often follow the same pattern of migration as camels in the semi-arid zone of the central belt (the Kababish and Hamar types), or the pattern followed by the Baggara cattle of the semi-humid zone.

POLICY AND LEGAL CONTEXT

Review of Land Law

As the prime source of wealth in the country, land has always been the focal point of Sudan's legal development.

> *At the very early stages and until now, customary law as applied in the different localities in the Sudan, legislation and Islamic law and precedents have regulated its ownership, user, possession, enjoyment and limitations on it.* (El-Mahadi,1969)

There are currently three forms of land tenure encountered in Sudan:

1. Absolute individual ownership: mainly along riverine areas where ownership is established by title of ownership and registration.
2. Overlordship or private landlordism: found over extensive areas of cultivable

land and pastures. This form of tenure was historically granted by rulers of indigenous kingdoms (that is, by Funj rulers) to tribal and religious leaders who have subsequently regulated the use of land by subjects, members or followers. Despite historical legal changes customary rights still prevail in many parts of the country, although to a limited extent.
3. Communal ownership: this form of ownership is vested in the tribal unit or the head of the tribe as 'overlord'.

The Condominium (a term used to describe the colonial government) government's land policy accepted these forms of land tenure, especially private ownership. The British rulers deemed all unregistered land to be government land according to the Land Settlement and Registration Ordinance of 1925. This states that 'all waste, forest, and unoccupied land shall be deemed to be the property of the government until the contrary is proved'. Within this framework of government ownership the British recognised customary tribal ownership, which in reality gave the right of use of land to tribal leaders only.

After independence the government challenged communal and tribal ownership when it started to establish large agricultural schemes on a tenancy basis and leased land to farmers which had hitherto been used by pastoral nomads. This alienation process has been legalised by a series of acts since the 1960s, culminating in the Unregistered Land Act of 1970 which gave the government power to transfer unregistered land to any public or national enterprise.

This act effectively deprived prior users from even the right to compensation or opportunities to be incorporated in the agricultural projects planned by the government. The 1970 act has worked to undermine pastoral production based on communal land ownership and has failed to produce an alternative form of land tenure to preserve rights of ownership or use. Tribal and communal land ownership is consequently becoming confined to areas of limited agricultural potential with no development infrastructure, and to areas normally reserved for conservation. Examples of extensive communal ownership are today only found in southern Sudan, mostly among the Nilotic pastoral communities.

Land tenure arrangements are determining the nature of use, allocation and management of land, and can be divided as follows:

- Areas of communal ownership: mostly under customary law and tradition, and mostly found in south Kordofan, Darfur and eastern estates, and unirrigated areas of some other states. Ownership here is understood as the right of cultivation, grazing, use of forest products, hunting and fishing, etc.
- Government land: based on the Unregistered Land Act of 1970. Allocated to individuals, farms or groups on long term lease for a specific purpose in con-

nection with irrigation schemes, mechanised rain-fed farming, and sugar plantations.
- Privately owned land: can be used for any purposes determined by the owner.

Pastoralists in Sudan use land within the concept of usufruct (*manfa'a*), which is defined as the right of using and enjoying land, and does not amount to ownership. The 1970 act clearly states 'a person in possession, use or enjoyment of waste, forest or unregistered land with or without express permission of the government shall be deemed to be usufructuary until the contrary is proved'. This provision works against pastoralists to whom land ownership is vested in the community, tribe or collective. All legislation and land tenure ordinances pertain to individual ownership at the expense of collective pastoral traditions. Even leasehold land granted by the government for periods of between 33 and 99 years only applies to largescale agricultural schemes on an individual tenure basis.

The Context of Pastoralism

The 1970 act has serious implications for pastoral communal land tenure, as it gives to central government and private investors legal access to land used by pastoralists that in the past was regulated by local government and native administrations. Although the native administration has been reinstated it does not have the strength to preserve pastoralists' rights under the government-supported privatisation process.

The act did not adequately define the legal status of current users, but gave government broad powers of eviction with complete discretion with regard to compensation. It provided the legal base for development projects which failed to incorporate traditional users, especially pastoralists whose customary rights had been taken care of by the native administration until its abolition under the terms of the 1971 Local Government Act. Until then the native administration had also been very effective in settling disputes, managing grazing resources and facilitating seasonal mobility.

The Regional Government Act of 1981, on which the present state structure is based, has given regional governments more power in land allocation and use, but remains subsidiary to the 1970 act, because national projects take priority. The 1970 act states that wasteland and forests, occupied or unoccupied, shall be the property of the government and registered as such.

One of the 1970 act's objectives was to promote national unity by not registering land in the name of a tribe or sub-tribe, although it did not reject customary claims to specific land. In this way it does not differ much from colonial legislation. Although priority is said to be given to local users as tenants in such projects, and every Sudanese national is entitled to a tenancy, nevertheless more outsiders have gained access to land and pastoralists have missed out due to their weak institutional and political power base.

The act has allowed the uncontrolled use of resources, increasing conflict between new and old users. Water resources (*hafirs*) provided by government with the understanding that access is free for all, have contributed to overexploitation of land. Under the tribal *dar*, these resources were regulated by the local authorities and annual tribal conferences. In the eastern region there are now over three million acres of mechanised farming recognised by the state which occupy the best pasture land. The struggle to demarcate a corridor of movement for pastoralists through the farming area is still not settled.

The Civil Transaction Act of 1984, which reflects Islamic *Sharia* laws, has complicated matters further. The problem is that different laws, including those relating to customary rights, are being applied at the same time but at different levels. This broad interpretation and application favours the acquisition of land by powerful interests and not the pastoralists, who have been marginalised by all land laws.

It is thus evident that legislation and policies related to land have been structured to favour individual, non-pastoralist interests, and are designed to meet the needs of individual owners of riverine land, and urban populations; commercial interests in leasehold tenure; and government and foreign investor joint enterprises which target high potential areas within pastoral lands.

The result is that legal changes severely disadvantage pastoralists—they can use land under the loose customary right of *manfa'a*, which confers the right of use of land owned by the government that is not claimed by anyone else, but once the government makes use of it or it is registered by an individual or group, the original user loses his *manfa'a* right unless he has used that land continuously for more than 20 years, in which case he will have priority to acquire that land. Pastoralists do not avail themselves of this provision, because they find it difficult to identify accurately the land they use. Consequently, they are being pushed off land they occupy and forced into more marginal areas.

This process reflects the misconceptions and prejudices that are commonly held about pastoralists: their way of life and production system is wasteful and unproductive and does not support the national economy. Pastoralism, however, contributes over 20 per cent of foreign earnings. In spite of this there is a persistent lack of integration of livestock with irrigated projects, which deprives pastoralists the right of access to land as herders, or as mixed crop/livestock farmers. A major constraint on more integrated development is the separation of institutions which concentrate on crop and livestock production. The Gezira model of cash crop production, a mixed farming production system dating from the British colonial period and based on a partnership between tenants and government, has been replicated in recent agricultural projects and has contributed to this division of crop and livestock production systems.

Pastoral land tenure is most often defined as the right to use communal lands generally unsuitable for crop farming, itself usually defined on the basis of private ownership or leasehold. Even forest lands are excluded by natural resource ordinances which prohibit the right of use by pastoralists except under very strict rules and regulations.

There is no specific national policy for livestock development nor is there a dedicated institution for this sector. This lack is one of the main reasons behind the marginalisation of pastoralists and the alienation of pastures. The extent of provision for pastoralism is confined to the opening of corridors for movement in north—south migrations to a limited extent in parts of Kordofan, and even less where mechanised farming is well established and the opening of such corridors would require the break up of officially registered schemes. These are defended by a strong and powerful mechanised farmers' union. The growing shortage of grain supplies has made the government less interested in supporting pastoral nomads in conflict with grain producers, who are politically and financially more powerful.

Other plans include provision of water sources in pasture land in the form of large *hafirs* designated as 'nomad *hafirs*' to be used by all pastoralists without restriction. This has complicated the situation of tenure and land use, as pastures around *hafirs* become overexploited. Traditionally pastoralists controlled pasture use through rights to water sources as individual or tribal wells, or a specifically assigned government *hafir* within tribal *dar* or territory.

Since the 1960s sedentarisation projects have been designed to serve pastoralists. However, such projects have failed to attain sustainability because pastoralists were not properly mobilised nor have they participated effectively in the planning of settlements. No attempt has been made to integrate their knowledge and expertise within the broader framework of the national strategy. These projects thus typify the top-down approach that is both impractical in its structure and its implementation. These highly localised projects have produced minimal impact on pastoralists or the environment.

TRENDS IN PASTORAL DEVELOPMENT

Development and Welfare

Trends in pastoral development in the country can be followed through various phases of development planning over the past three decades. Service programmes and water provision, especially in the late 1960s and early 1970s, were not preceded by a proper resource survey, livestock count or assessment of pastoral migration. During this period, according to the annual report of a rural water corporation, the government was committed to tackling the severe water shortage in rural areas. In spite of this, predictable overstocking around water points, as well as overgrazing and conflict with settled farming populations has occurred.

From the 1950s to the early 1980s a priority of the government programme was to provide water for both people and animals. It started by drilling boreholes and constructing *hafir*. The *hafir* programme is concentrated in the savanna country, south of the Nubian sandstones, with its cracking clay soils which are suitable for *hafir* construction. *Hafir* were also constructed without proper planning which has similarly resulted in land degradation.

This form of development has had two adverse effects on pastoralists' welfare. First, as the new water sources were open to everybody, this complicated the use of grazing resources and caused overstocking. Second, improvements in the water supply were not coupled with a comparable improvement in range and pasture quality. Together these made the pastoral production system more vulnerable to climatic variables. Average livestock losses were estimated to be 60 per cent during the 1984 drought. During this same period, pastoralist settlement projects were attempted in different areas in the country, but with the same outcome.

In the 1970s and 1980s, development projects concentrated on high potential areas and largescale government schemes. The pastoral sector received no substantial attention as it was thought to be of low potential and less likely to provide substantial returns on investment. At the same time it also lost grazing land to agricultural schemes.

The severe economic recession has now halted all development, especially in the pastoralist sector. As a result pastoralists have been severely affected by drought and desertification, and are rapidly losing their livestock. Meanwhile, the government is more concerned with solving the problem of food shortages and famine rather than ensuring that pastoralism continues as a means of survival for a significant proportion of the population. Neither does pastoralist development appeal to donors because it is not thought to provide a quick return on investment in the manner demanded by foreign funders.

In the current period of limited development in the country there are isolated rehabilitation activities: the UN is active in southern Obeid, Umm Kaddada, Central Butana and Lower Atbara, and OXFAM and ACORD (Agency for Cooperation and Research in Development) also operate projects. The Department of Range and Pasture Management, in spite of its limited resources, is engaged in the improvement of range by enclosing rangeland and employing pasture improvement technologies. These include the rehabilitation of pastures around El Odiya in central Kordofan, production of fodder in the eastern state, opening fire-lines, rehabilitation of Hamrat el Wiz range, and seed spreading in Darfur.

In the south development has stopped because of the high level of insecurity caused by the war. Even during the brief years of peace since independence no attempt has been made to support pastoral development. Some initiatives were taken by the Jonglei Canal Development, but this was stopped ten years ago. The war has disturbed traditional sustainability and resulted in food shortages and the displacement of populations, with out-migrations and the near collapse of pastoralism, especially among the Dinka—the largest livestock breeders in the country.

Development in the non-pastoral sector relates to activities in mechanised and irrigated agriculture. This has compounded problems for pastoralists by restricting access to pasture land.

The completion of the Nyala-Omdurman stock route has accelerated livestock transfers from nomads to traders, exporters and dealers. Farmers of all types have gained the advantage of owning land and cultivating fodder while pastoralists only have access to natural pastures. Farmers also have access to credit to support their activities.

Over the past decade a major process of wealth transfer has taken place among pastoralists. Because of severe drought, people were forced to get rid of their animals in order to meet basic needs. Owners and tenants of large farms were the main buyers. The backing of unions enabled them to use the legal system to their advantage. As a result more livestock is now being bred by agro-pastoralists and more pastoralists are being forced out of pastoralism, migrating to urban centres or to modern agricultural projects. In many cases women support the entire family by working as house servants in towns and cities.

MAIN AREAS OF CONFLICT

There are three types of land conflicts and disputes in the Sudan:

1. between pastoralists competing for pastures, water, and cultivable land, especially in the dry season and in years of severe shortage;
2. between pastoralists and settled populations who have extended their farming activities onto grazing land or who have cut off livestock corridors;
3. between pastoralists and largescale agricultural projects established on land customarily claimed by pastoralists.

Such projects also include forest reserves and sugar plantations, which are structured in such a way that they prevent pastoralists from having access to land they would traditionally regard as rightfully theirs.

Conflict is becoming more common due to the rapid degradation of rangeland areas. This has forced pastoralists to extend their movement southwards, where agricultural activities and settled populations are more concentrated. Before British rule, disputes between pastoralists and traditional farmers were settled either by warfare or by peace agreements made by tribal leaders. Acquisition of land through war structured the 'tribal *dar* map', which was recognised by the British and formed the basis for local government. The British also recognised the tribal leadership hierarchy as a base for native administration, and so incorporated it into their administrative system.

The British effectively pre-empted many land tenure conflicts by granting legal powers to native administrations. They appointed tribal leaders who were supervised by administrators to arrange for pastoral movements and settle land disputes. The Conference of Tribal Leaders made decisions that were given the power of law by the government. Sudan's national archives have a wealth of proceedings from such conferences and laws based on them. For instance, the governor of Kassala Province wrote in 1928:

In reference to development of agriculture in Butana, and in view of the possibility of future development and of the attempt to establish 'dar' right in the area, I think it would be as well to put on record the right claimed by Butana tribes.

Based on another meeting the governor wrote:

Hadendowa and Beni Amir cultivating south of Gargaf pay their Ushur (crop taxes) to Nazir of Shukriya. In conclusion, if however it is at any time proposed to open the area up by digging wells, increasing cultivation or by making any permanent village, I hope the office may be informed before action is taken, as such action might be prejudicial to the rights or welfare of Butana tribes and in particular of Lahawin. (Sudan National Archives, Red Sea Province Collection unpublished)

The same effective system of conflict resolution was extended in the early years of independence. In June 1958 the governor of Darfur instructed the southern Darfur District Commissioner to allow free movement of tribes, but this did not mean challenging the previous agreements which ensured that no tribe could take another tribe's land. The governor instructed Zalengi and Nyala rural council officers to issue ordinances to confirm these directives.

Involving pastoralists in conflict resolution had been the norm up to the 1970 and 1971 acts which abolished the native administration. Since then the government has resolved land conflicts with a procedure that excludes pastoralists. This has been based on the belief that the government is best able to see what is good for the people. Neither an effective mechanism nor institutions have replaced the native administration and pastoralists have been left with a new legal system about which they know little and which they consider to be biased against them. Frustration is compounded when the government allocates land to people the pastoralists consider to be outsiders with no customary rights to the land.

This has caused tribal conflicts in many cases. For instance, the West Savanna Project was partly established on Ma'alya tribal land which had been allocated to the Rezeigat. When the project failed the Ma'alya reclaimed their land, but the Rezeigat refused to recognise this on the grounds that the land belonged to the government. The result was bloody armed clashes between the two tribes.

Tenancy leasehold in New Halfa, Rahad, and similar projects, is granted to persons who are settled, with previous experience in agriculture and willing to stay permanently in the scheme and abide by the ordered cropping system. Such conditions are satisfied by very few nomads. The pastoralists' need for access to water and crop residues in such projects during drought years has provoked new conflicts with the government. In the belief that pastoralists cause crop damage, they are kept away by police and military forces, and in the case of damage they are subjected to severe punishment by law. Having no trust in the courts, they opt for paying bribes to police who take advantage of this and victimise them even when there is no evidence that they have caused damage. In mechanised farming areas, scheme owners employ armed guards and pay for uniformed police to keep pastoralists out. Incidents of guards shooting pastoralists and their animals have been reported to police and taken to the courts in Gedaref District.

In many cases conflicts are settled by police outside the courts. Tenants and farmers are generally backed by their strong, officially recognised unions, while pastoralists are not so organised and lack the knowledge and means to make use of the legal system. The succession of military governments and resulting changes in laws have made it difficult for pastoralists to cope with the many, and often inconsistent, land laws.

ENVIRONMENTAL IMPLICATIONS

The contribution of pastoralists to environmental degradation in Sudan and elsewhere is a very controversial issue. Many argue that through their variable and selective use of resources, pastoralists provide the highest utility of rangeland resources, particularly where space and freedom of movement is found. The counter argument is that communal tenure does not encourage conservation.

Although pastoralists maintained a kind of equilibrium with the environment in the past, they now can no longer do so. This is reflected in environmental degradation in pastoralists' homelands (Map 20). The pastoralists are blamed for the degradation through the apparent lack of control over livestock numbers and their perceived economic irrationality. This explains the push for settlement of pastoralist communities and for the type of development projects originally planned for pastoralists in the 1960s and 1970s.

However, when comparing pastoralists with settled communities sharing the same area of land and resources, the intensity of land use by agricultural settlers poses a more serious process of degradation. The intensive use of soil by settlers using traditional means with minimum input, the demand for woodfuel and building materials, and livestock's need for fodder, are even more damaging to the environment than the pastoralists' traditional activities. Mechanised rain-fed

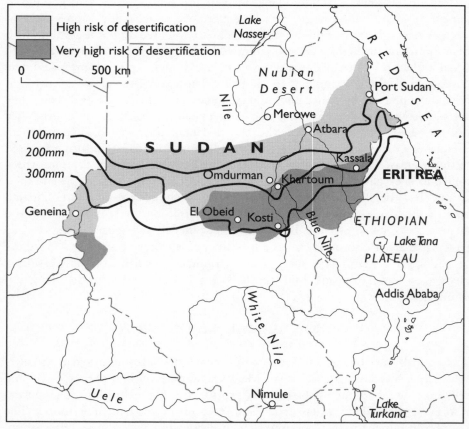

Map 20 Environmental Degradation in Pastoral Homeland Sudan

agriculture depletes soil fertility and does more harm than grazing. Urban populations also affect the rangelands in their demand for energy, building materials and food.

Because mobility as an effective form of resource use is no longer possible due to shortage of land, pastoralism has proved to be unsustainable. Therefore, pastoralists have had to adapt accordingly. Pastoralists are left to 'scratch' for meagre resources, and their impact on the environment appears to be becoming more serious compared with other land users. This environmental impact is associated with the following main factors:

Water

The provision of water in pastoral areas without sufficient planning and better assessment of resource potentials has resulted in overstocking around water points. This has concentrated livestock in ecologically marginal zones. Pastoral mobility has been disrupted and pastoralism has now extended into the better areas to the south where competition for land is more acute under open access grazing pressure.

The concept of the 'tribal *dar*' has been abolished and indigenous mechanisms of resource use control have become ineffective. Under such conditions the general impression is that pastoralists are contributing to the degradation of the central belt of Sudan.

Drought

Successive years of drought have reduced resource potentials dramatically. Although drought has reduced livestock numbers, they still exceed the potential of available resources. In spite of the losses caused by drought between 1983 and 1990, the rapid recovery of small animals (sheep and goats) has been enough to increase stocking rates on degraded lands. Many pastoralists make use of purchased or cultivated fodder which has reduced the losses of stock during droughts.

Land

The area of land available for sustainable pastoralism based on mobility has been reduced as a result of persistent droughts and the expansion in farming, which is not structured to accept livestock as part of the production system and thus denies pastoralists access to vast tracts of land by law. For example, see Maps 21 and 22 for comparison of the pastoral environment in Blue Nile Province between 1955 and 1986.

The impact of these factors is manifest in the small amount and poor quality of pastures, reduced carrying capacity and soil degradation, and a decline in pastoralists' quality of life. Pastoralists have responded to adverse conditions by adopting various strategies; for instance, through acquiring tenancy of plots on irrigation projects in their traditional homelands through the bureaucratic acquisition of leases, the buying out of existing leases, and sharecropping with the objective of having legal access to water sources and crop residues.

Pastoralists have also formed alliances and provided services to rich herders—agro-pastoralists who have access to crop residues and water sources constructed by scheme owners who have political influence. In addition they have affiliated with and become incorporated into rehabilitation and development projects sponsored by the UN and NGOs with the aim of gaining access to land, diversifying their economic activities and linking up with the market economy.

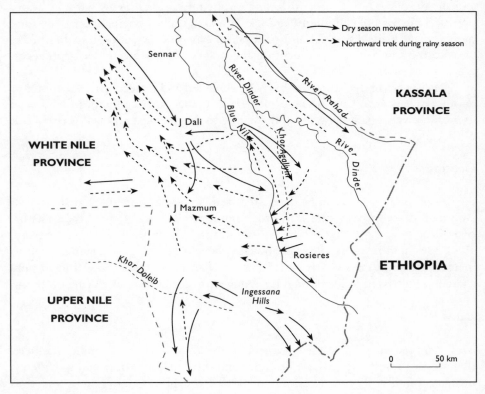

Map 21 Nomadism in Blue Nile Province, 1955

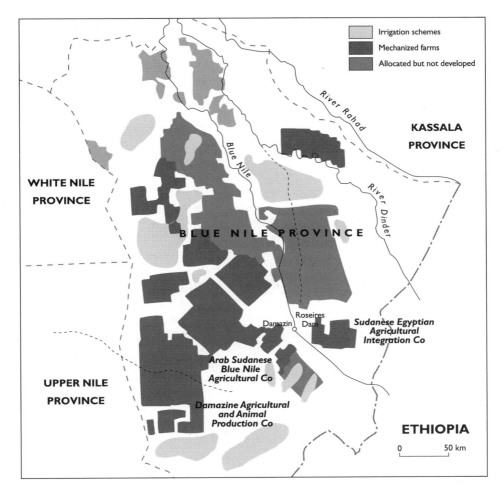

Map 22 Agricultural Schemes in Blue Nile Province, 1986

The increasing failure of pastoralists to avert a crisis is seen in their inability to satisfy basic human needs, the inability of the pastoral system to deal with drought and famine, demographic instability, the high rate of pastoral 'dropouts', and the shift from degraded pastoralism to urban poverty.

CASE MATERIAL

Pastoral land tenure is the least considered dimension in published studies on Sudan, and such studies as exist are generally poor. Land tenure is simply accepted as being communal or tribal, without more thorough examination.

Traditional pastoral land use practices and changes occurring to these systems are issues of academic interest, and have been documented by research institutions. The University of Khartoum, the largest academic institution in the country, has a reasonably large collection of data on pastoralists, but it is highly fragmented and reflects the different interests of researchers and their strong leaning towards particular disciplines. Only a few cases have a cross-disciplinary approach.

The dimensions of pastoral production and its contribution to the national economy have never been studied. Economic surveys and official annual statistics define the contribution of livestock to the national economy in terms of local supply and foreign earnings, but omit any cost-benefit analysis of this production, almost 80 per cent of which is derived from the pastoralist sector. Without scientific studies it is difficult to assess the long-term viability of pastoral production. Pastoralism has never been assessed in comparison with other uses of pasture land, particularly mechanised and irrigated farming which, unlike the former, receives no subsidy from the development budget. The ability of pastoralism to support pastoral populations remain uncharted.

The lack of scientific analysis of customary land tenure arrangements among pastoralists in Sudan is one of the main reasons why the government has been able to disregard it as a productive form of land use. Most of the legal changes in the country are based on the premise that customary land tenure hinders development. It also explains the interest in sedentarisation as a means of strengthening attachment to land and minimising the perceived waste and misuse of resources under communal tenure systems.

The initial research interest in pastoralism came from anthropologists who were more interested in pastoralists as a distinct cultural group. Geographers and economists also avoid addressing land tenure issues and concentrate on the pastoralists' production system (Lebon 1965).

The changes in pastoral areas have attracted more attention from researchers, NGOs, and government agents. The issue of land tenure is almost completely ignored by development planners and decision-makers. Pastoralists, like some other groups in Sudan, are victims of a professional bias that separates technical and social disciplines. Technicians consider the humanities as academic and not applicable to development planning. Recently, with the development in behavioural sciences and the realisation of the value of an integrated approach to development, a new tolerance has emerged in academic circles. But this is not yet shared by planners and decision-makers.

PROCEDURES FOR CONFLICT RESOLUTION

The present government has reinstated native administrations and advocated a federal system of administration in the country. This is expected to provide a more suitable medium for the involvement of pastoralists in land disputes. Although the native administration has not yet been granted full power, it has provided pastoralists with a system to resolve disputes more effectively.

The system of native administration, however, is overshadowed by institutions formed to provide support for the political establishment. Pastoralists do not form an obvious, reliable body of support for political regimes, so they are not a focus of attention compared with urban and rural settled communities who use political tactics to gain the notice of the government.

The lack of access pastoralists have to political and legal systems is matched by the lack of knowledge and decision-making skills in the pastoral communities. In November 1991, a land dispute emerged between the Beja sub-tribes over the use of land flooded by the *khor Goub* (seasonal water course) after the construction of an earth embankment by the Soil Conservation Department. The Tokar commissioner threatened to distribute the land in the way he saw fit. He was advised during the height of the dispute, which drew people from as far as Gash, to try and use the local tribal leadership, which he did. With their help the dispute was resolved within a day.

The OXFAM integrated model of development in north Tokar has been successful in handling the complex issue of Beja land tenure. OXFAM encouraged people to settle land disputes before receiving its aid. The programme is a community-based development venture that encourages pastoralists to settle land disputes themselves. It also mobilises them to seek government recognition of their rights to land, very much like leasehold tenancies.

Similar programmes are being started by the UN in selected areas, for instance, south Obeid (Kordofan), Idd el-Ghanam, Umm Kaddada (Darfur), Central Butana and Lower Atbara. The main advantage of these programmes is the identification of specific pastoral areas where development is taking place. They also protect them from being earmarked for non-pastoral development. They are based on local participation, and focus on building institutions which defend pastoralists' land rights by exerting pressure on the government. In such projects pastoralists claim similar rights to those held by tenants on irrigated schemes and mechanised farming scheme owners.

The Range and Pasture Department, in its rehabilitation schemes supported by international organisations, has opted for this same model which involves local people in all stages of planning and implementation, for instance, the El Odaya project in Kordofan.

SUMMARY, CONCLUSION AND RECOMMENDATIONS

Trends in land tenure in Sudan disadvantage pastoralists. Rangeland is being ceded to agriculture, and it is obvious that an anti-pastoralist bias in government is an important factor in the decline of pastoralism. It is also evident that pastoralists are the least advantaged and the poorest competitors in the race between government, private investors and traditional farmers. The government is not only the legislator, but also the main user of land. Pastoralists, whose political influence is very limited, are being displaced from the land they traditionally used and forced onto more marginal land, to which they have limited rights of ownership, and which they have neither the power nor the means to secure.

Land tenure analyses and the impact of tenure on pastoral production systems represent a real gap in research. Much of the published data stress the fact that pastoralism has internal survival mechanisms, but no serious work has been done on how land tenure operates to achieve this. The data emphasise social networks, mobility and the role of pastoralism in the national economy as the justification for top-down development to replace pastoralism with something more 'productive'.

The main conclusions of this study are:

1. Land legislation disadvantages pastoralists by overturning customary laws in favour of new legislative provisions. The 1970 act provides the main threat to the pastoral land tenure security instead of protecting pastoralists by either accepting customary rights or devising a legal framework to observe their rights. Under the act the government has allocated land for what are called 'national projects' which favour settled populations and big investors. The trend towards 'spearhead development' targets areas and communities with high potential and limits instead of enhancing pastoralists' opportunities in such developments.
2. Pastoralists have not only lost their land but also, with the abolition of the native administration, lost an important official link with government. Although it was reinstated in 1989 its role is marginal as its structure and mandate are limited.
3. Pastoralism has been evolving to cope with the changing conditions. But current land tenure arrangements have undermined the potential of the new strategies. This means increasing vulnerability to adverse environmental conditions, loss of wealth, food shortages, the rejection of pastoralism as a way of

life, destitution and marginalisation of pastoralists.
4. Research on pastoralists was initially confined to studies of pastoralist cultures. In the 1980s pastoral systems were examined for their role in 'desertification'. Since then research has become more interdisciplinary in the service of development projects. However, most of it remains mostly academic, fragmented and inaccessible to decision-makers who approach the issue from an urban background.
5. Conflict resolution procedures are unfair to pastoralists. In legal disputes they are obliged to cede their rights to others who have the support of the government.

The main recommendations of this study are:

1. More research on pastoral land tenure systems and how they affect pastoralism's development is needed. This should be formalised, interdisciplinary in nature, and focused on specifics if it is to avoid previous failures. It must involve government institutions at all stages, and bring understanding of the effects of existing tenure arrangements, development processes, and their economic and environmental impacts on pastoralists.
2. Alternative modes of development and assessment of participative procedures need to be researched. Pastoralists' institutions should be researched as a follow-up to the NGO and UN programmes in the country.
3. A pastoralists' research centre that explores all relevant issues and provides the scientific basis for decision-making should be established as a means of achieving these objectives.

7

TANZANIA

D K Ndagala
Ministry of Education and Culture[1]

Map 23 Tanzania

INTRODUCTION

Tanzania has 13 million head of cattle, the second largest national cattle herd in Africa. These are almost wholly owned by pastoral and agro-pastoral societies, most of which are found in the areas indicated on Map 24. The contribution of the livestock sector to GDP is derived mainly from pastoral smallholder herds.

Any attempts to increase or control the products of the national herd are, implicitly, attempts to increase or control pastoral production.

Pastoralism consists of an interaction between herders, animals and a given mode of resource management (Mohamed Salih 1990). It is, potentially, an economically viable and sustainable form of land use (Lane and Scoones 1991). Communities dependent on livestock range from those which practise little or no agriculture, to agro-pastoralists that do, but have a strong economic and cultural leaning towards livestock. The most pastoral communities in Tanzania are the Maasai, the Ilparakuyo, and the Datoga, of whom the Barabaig are the largest section.

In the 1980s pastoral areas were subjected to a wave of stock raiding which left hundreds of people dead and thousands of cattle and small stock unaccounted for (Ndagala 1991b). During the country's economic crisis in the same period, many animals died of disease because of the lack of dipping facilities. The main concern, however, continues to be the decrease in available pastures because of growing pressure on pastoral land. Despite the wealth of research data on this issue since the early 1980s (Galaty and Aronson 1981) this concern has not yet been addressed. Nevertheless, a consensus has been reached that the shrinkage of pastoral land in East Africa is connected with the issue of 'tenure'. During a recent workshop in Nairobi, it was realised that more data were still needed and that a research agenda had to be drawn up to facilitate this (Kituyi and Kipuri 1991).

LEGAL CONTEXT

Before colonialism all land was held under customary tenure. Most Tanzanians still use land according to customary laws. The principle behind this is clearly explained by James and Fimbo (1976):

> *In Africa, and particularly in Tanzania, land is more than property, and land tenure rules form part of the whole complex of culture. Land tenure rules are dominated by the need to protect groups of specific parcels of land – each unit whether it is a family, clan or wider community has a corpus of law which is directed against outsiders and defining the rights and duties of all within the group.*

In spite of the important variations which exist, generally an individual obtains land rights as a member of a community. Mwalimu Nyerere, Tanzania's first president, explained it in this way:

> *To us in Africa land was always recognised as belonging to the Community. Each*

Map 24 Area of Major Land Alienation in Tanzania's Pastoral Areas

individual within our society had a right to the use of land, because otherwise he could not earn his living and one cannot have the right to life without also having the right to some means of maintaining life. But the African's right to land was simply the right to use it: he could have no other right to it, nor did it occur to him to try and claim one.

Things began to change after German occupation in the late 1890s. Many areas with potential commercial value were given to European settlers for definite periods with options to purchase, and leaseholds were given for indefinite periods.

This move was facilitated by the German Imperial Ordinance (GIO) of 26 November 1895. Accordingly, except where claims to ownership and to real rights in land could be proved by individuals and certain designated persons, all land was to be deemed unowned and to be regarded as belonging to the Crown. Ownership to such land was vested in the *Reich*. Existing rights of individuals were to be recognised by specially designated commissions, and proof of a title in land was generally to be by documentary evidence. Under customary land tenure Africans had no documentary titles to their holdings, and as their occupation was deemed permissive, they had no security of title and lost their rights (James and Fimbo 1973).

Although Africans continued to use their land according to their customs after enactment of the GIO, their land was no longer protected from indiscriminate alienation by the colonial administration.

After the First World War, the British pursued similar policies. In 1923 they passed the Land Ordinance which paved the way for further alienation of native lands by giving rights of occupancy to non-indigenous peoples. Section 4 of the Ordinance stated that:

> ...all native lands and all rights over the same are hereby declared to be under the control and subject to the disposition of the Governor and shall be held and administered for the use and common benefit of the natives of the Territory, and no title to the occupation and use of any such lands shall be valid without the consent of the Governor.

Land rights were a means to a livelihood for native peoples, which the Ordinance effectively placed under the control of the governor. This was an important turning point in resource control.

Customary land tenure was recognised in the Land Ordinance of 1928, which gave the right of occupancy to those 'using or occupying land in accordance with native law and custom'. However, this was watered down in 1947 by subsection (2) of Chapter 113 of the 'Laws of Tanganyika', which declared that all land was public, and so placed Africans at the mercy of the colonial administration. It was the administration which could determine whether the rights in a given piece of land were lawfully acquired. Though this ordinance was amended in 1950 to provide for consultation with native authorities before land occupied by Africans was allocated to settlers, this provision was often disregarded. Many administrators interpreted the requirements of consultation to be discretionary and, in some instances, dispossessed Africans were not compensated for improvements made to the land (James and Fimbo 1973).

Though land alienation affected many native communities, pastoral communities seem to have been the worst affected (Rigby 1985). Even in places where the colonial administrations showed restraint in alienating native land, pastoral land was often assumed to be unowned or unused. The whole concept of land use was (and continues to be) tied to the displacement of grass by agricultural crops and constructions. The rights of land pastoralists have therefore remained vulnerable because pastoralism provides little evidence of such displacement.

The rights of Africans were encroached on to make land accessible to non-Africans. And by not insisting on user-rights, as is the case under customary land tenure, the new legal provisions enabled non-Africans to become absentee owners of land. But at the time of independence the leaders of the new state were anxious to tackle the land question. This was inevitable because nearly 98 per cent of the total population was rural and dependent on land as its main source of livelihood. Many of the Africans who were dispossessed of their land during the colonial period revived their grievances after independence. The Land (Settlement of Disputes) Act 1963 attempted to resolve these grievances. In the same year, however, the Native Authorities Ordinance (Repeal) Act abolished all the remaining native authorities. As a result, the structure of traditional hierarchy in relation to land administration crumbled, and the method by which land held under customary tenure could be adjudicated remained obscure (James and Fimbo 1973).

Mwalimu Nyerere's ideas greatly influenced the policies of the new government immediately after the country became independent. His thinking seems to have been a combination of African traditionalist and socialist ideology, informed by the need to unite the different ethnic groups which constituted the young nation of Tanganyika. He was against individual ownership of land, and directed that:

> *The Tanganyika Africa National Union (TANU) government [Tanganyika and later Tanzania's sole ruling party] must go back to the traditional African custom of land holdings. That is to say, a member of society will be entitled to a piece of land on condition that he uses it. Unconditional, or 'freehold' ownership of land (which leads to speculation and parasitism) must be abolished.* (Nyerere 1966)

In saying this, however, he was not suggesting that natives of a given area could lay claim to land on the grounds that it belonged to them by custom. Rather:

> *In the past, when our population was divided into different tribal groups, the land belonged to the particular tribe living on it. In future, however, our population will be united as one nation, and the land will belong to the nation. And today just as one man cannot prevent another man from his tribe from using land, so also tomorrow one tribe will not be able to prevent another tribe from using land that is actually the property of the nation as a whole. Our aim is to reach an arrangement for distributing land which we can use to meet our requirement.* (ibid)

Not only was the right of different tribal groups to control land implicitly abolished; the possibility of reserving land for particular communities was also ruled out:

> *What we must not be asked to do—and it would be absolutely dishonest for us to say we could do such a thing—is to reserve any particular area of land for the Asian Community, or for the European Community or any other community. That we cannot and will not do.* (ibid)

Individual ownership of land was rejected, but the individual use of land by way of leasehold agreements was approved. Leaseholders could use land as they wished, for example, to raise a loan. But leaseholders soon came into conflict with many rural people, for as more people took leases many others were turned into 'squatters' on the land they had used all their life. The government's solution was to pass the Rural Farmlands Act in 1966, in which a measure of compensation was guaranteed to those who had used land before its acquisition by leaseholders.

As a result of these policies and legislation, land was transformed from a communally owned resource into a public resource controlled by the state (Ndagala 1991b). Individuals or groups had user-rights which could be held as long as the state deemed it fit. People from one part of the country could have access to land in other parts of the country. This move was hailed as a way of consolidating unity among citizens. Although in principle pastoralists and agriculturists alike could leave their customary lands to live elsewhere in the country, the dominant trend was for cultivators to move into pastoral areas.

Prejudice Against Pastoralists

Pastoralism in Tanzania, and indeed in the whole of East Africa, is widely viewed as economically unsound. Despite attempts to fight this sort of thinking (Rigby 1969, Raikes 1981, Ndagala 1974, 1986, 1990a, 1991a, Mustafa 1989, Lane 1996), the misconceptions and prejudices still prevail. Pastoralists are said to accumulate cattle for social prestige rather than economic need (Herskovits 1926). Moreover, pastoralists are said to wander at random (Raikes 1981), thereby making it difficult for governments to provide them with basic social services. This view has been the basis for pastoral settlement programmes, such as Operation Imparnati in Maasailand (Ndagala 1982, Parkipuny 1979) and Operation Barabaig (Ndagala 1978, 1990c, 1991b, Loiske 1990).

The misconception that pastoralists wander randomly gives rise to the belief that pastoral claims to particular land are fluid and temporary. This, and also the supposition that land not grazed at any one time is 'free', have resulted in the pastoralists losing a great deal of land without receiving compensation. Such losses are reported among the Ilparakuyo (Ndagala 1974, 1986, 1990c, Mustafa 1989), the Barabaig (Lane 1990a, 1990b, 1993, 1996, Lane and Pretty 1990, Lane and Swift 1989), and the Maasai (Parkipuny 1977, 1979, Arhem 1985a, 1985b, Ndagala 1990a, 1990b).

Pastoral production is considered by many policy-makers to be backward in relation to farming. The immigration of farmers into pastoral areas, where relatively high potential grazing land has been put under cultivation, has been regarded as a 'civilizing mission' (Raikes 1981). In the process of making 'unused' land productive, the farmers are believed to soften pastoral resistance and expose pastoralists to modern ways of living. In this way, many thousands of acres of grazing land have been lost to farming in all pastoral areas (Arhem 1985a, Loiske 1990).

Communal land tenure, under which most pastoral land is held, has been under continuous attack since colonial times. In 1948, for instance, the director of Veterinary Services wrote:

There is one overriding stumbling block that is the system of uncontrolled communal land tenure which permits of the fiercest competition taking place for every blade of grass and every drop of water. ... pasturage, the life-blood of animal husbandry, is the common property of all and consequently little effort is made to improve or indeed preserve it.

In 1989, 40 years later, an official of the Tanzanian Ministry of Agriculture and Livestock Development echoed this argument:

The practice of grazing private livestock on communal land constitutes the single major constraint to improved management of the natural pasture lands. The inevitable result of this system of livestock production is that the cattle owners keep excessive numbers of livestock which in turn leads to overgrazing, soil degradation, low fertility and high mortality rates. (Bilali 1989)

However, communal tenure remains the main system under which pastoral land is utilised (Tanzania Government 1982). Little effort has been made to establish the merits of this system. Planners unthinkingly accept the 'tragedy of the commons' thesis (Hardin 1968), which claims that individual herders have no incentive to restrict stock numbers, and that their herding of private animals on communal pastures will inevitably lead to overgrazing and land degradation.

Communal land tenure has been identified as the major obstacle to natural resource management and livestock development in many African societies. The usual technical solution proposed to this problem is privatisation of land through the introduction of individual tenure or various forms of group tenure. (Helland 1990)

In spite of the many anthropological criticisms of Hardin's argument, and Hardin's own recent re-formulation (Hardin 1986) the old orthodoxy continues to be espoused (Lane and Scoones 1996).

Nevertheless, the Tanzanian government has made two major attempts to grant communal rights of occupancy. The first was in 1964 with the Range

Development Act. This was immediately applied to the whole of Tanzania Maasailand (the present Ngorongoro, Monduli, Kiteto and Simaryiro districts, excluding the Ngorongoro Conservation Area), where it was declared a Range Development Area under the Maasai Range Commission. The commission's role was to mobilise people to form ranching associations to which land was allocated. Once this happened, all customary rights were extinguished. The association was automatically granted the power to enter into normal commercial dealings with the land or part thereof. A member of a ranching association was entitled to reside on the ranchlands of the association together with members of his household; to keep and graze stock on the ranchlands, not exceeding a set quota; and to such other rights of pasturage, water, cultivation and enjoyment of the natural resources as may be provided for in the rules of the association.

Of the 22 ranching associations which were to have been formed, only eight had been registered by 1980. Before the steering committees had gained the necessary competence to run their associations the whole exercise was overtaken by the nationwide 'villagisation' programme which established several villages in each of the areas earmarked for ranching associations, most of which were later registered under the Villages and Ujamaa Villages Act of 1975. The registered villages were to be given rights of occupancy by the government to provide them with security of tenure in land falling under their jurisdiction. This was the second attempt by the government to promote communal group rights of occupancy. However, up to now most of these villages have not been given rights of occupancy. This is because the villagisation programme stopped where it should have started (Ndagala 1990a, 1990b)—the permanent settlements established during the programme were to have been nuclei of viable economic units with clearly defined landholdings that were legally protected from encroachment, but the settlements have remained ends in themselves whose residents continue to lose land through encroachment by outsiders and alienation by the government.

Immediately relevant to pastoral land tenure is 'The Livestock Policy of Tanzania' (Tanzania Government 1982). Whereas promising attempts were made to give legal security to communal landholdings through the formation of ranching associations and villages, the livestock policy gives no consideration to these possibilities. Instead it supports the view that private landholdings encourage gainful investments of effort and money (Lane 1990a). Pastoralism which, together with agro-pastoralism, constitutes what the policy calls the 'traditional sub-sector', is regarded as backward. Its land tenure system of communal grazing is said to lead to overstocking, overgrazing and destruction of soil structure.

The thrust of the policy is thus to transform the backward traditional sub-sector into a modern sub-sector. One of the problems facing the transformation is said to be the 'traditional producers' attitudes and practices' (Tanzania Government

1982). It is proposed that since it would take a long time to change those attitudes, emphasis in the short term should be given to expansion of the commercial sector. This means that the traditional producers, who keep over 99 per cent of cattle, goats and sheep, would have to be side-stepped in favour of the commercial sector, which accounts for less than 1 per cent of all livestock. This livestock relegation of the traditional sector has nothing to do with the producers' attitudes but arises from bourgeois modernisation theory (Mustafa 1989). For example, it is stated that:

Even though the majority of milk produced in Tanzania comes from the traditional herd and even though there is potential for increasing production from this source, the greatest emphasis will be placed on expanding the size and increase in the productivity of the grade dairy herd. (Tanzania Government 1982)

The implications of this policy are that large areas held under customary land tenure would be allocated to individuals and parastatals for commercial ranching or dairy farming. Though the contribution of these new ventures to overall livestock production is still minimal, their impact on resource availability is considerable. Livestock production and resource sustainability are unlikely to be improved without the involvement and cooperation of traditional livestock keepers.

TRENDS IN PASTORAL DEVELOPMENT AND WELFARE

What has been taking place on the rangelands in Tanzania is partly a result of the policies and prejudices outlined in the preceding section, but also partly because some of the projects undertaken on the rangelands have had nothing to do with the policies outlined above. The installation of water and dipping facilities, for example, has in many instances been a positive response to the perceived needs of the people as presented to the government. Nevertheless, there are several examples of failed projects which are a direct result of the inappropriate policies described above.

In Arusha Region, the Maasai Range Project was established on the assumption that Maasai herds were poorly managed, poorly nourished, diseased and underharvested, and that the only economic way to manage the extensive rangelands was to manipulate the existing features of the environment and maximise production within the longer term constraint of protecting its ecological potential (Morris 1975, Ndagala 1990b). It is not clear yet whether the relative lack of interest in giving rights of occupancy to the villages in Maasailand has something to do with the general official prejudice against communal land tenure.

Up to now most of the villages have not been surveyed to establish the area they need to be viable. There are mounting claims that the large areas which should in fact belong to the respective villages are unwittingly leased by district councils to commercial farmers on the grounds that they cannot be left idle when somebody else needs them for productive purposes. The establishment of wheat farms in Hanang District whereby high potential Barabaig land was alienated (Lane 1990a, 1990b, Lane 1996, Loiske 1990, Ndagala 1990c) and the licensing of largescale commercial farms such as those owned or backed by the Arusha-based Dutch seed bean companies (Kjaerby 1979, Loiske 1990) are both examples of the official misconception that the land was free and had to be put to better use—that is, farming.

In Bagamoyo District, state farms and ranches were established on high potential pastures to the dismay of many pastoral households. For example, 44,000 acres were alienated for the extension of Ruvu Ranch. The area was said by Ruvu Ranch officials to be underused by the resident 276 pastoral households. Yet those families had 7044 head of cattle, not to mention small stock. At a stocking rate of six to 12 acres per livestock unit, which is the average for that area, the land in question was just enough for that number of animals and, therefore, fully utilised by the resident pastoral households. Yet the households lost their claims to the Ruvu Ranch (Ndagala 1974, 1986, 1990b).

Livestock development programmes have mainly focused on pastoralist herds —especially in the northern parts—in order to increase offtake, add to the food and income supply, and also to control herd size to reduce overgrazing (USAID 1986). The programmes which have had most impact on land tenure are the Phase I and II World Bank programmes and the USAID-supported Maasai Range Project. Among the projects that were to be supported under the World Bank programme were development of 11 National Ranching Company (NARCO) ranches, four District Development Corporation (DDC) ranches, and 22 Ujamaa cooperative ranches.

NARCO and DDC ranches constituted the state ranches to which pastoralists had little or no access. The amount of land under their control is large. For example, the DDC ranches averaged around 40,000ha. The formation of NARCO ranches was:

...a direct result of President Nyerere's publicly stated belief that government-operated facilities would be needed to supply cattle for export, tourism, and also for critical food needs during a period of national transformation. (ibid)

The Ujamaa ranches, on the other hand, were established by volunteers from the respective villages. These ranches were used more as holding areas, and the herds manipulated by their former appropriating 'owners' despite official designation of the herds as 'communally owned' (ibid). Moreover, nearly every one of the Ujamaa ranches had a high density of livestock and people. This is because as long as these ranches were operational they would not be subject to land alienation. Therefore, to the pastoralists who had lost plenty of land to state ranches and farms, the Ujamaa ranches were the best alternative resources available to them.

What has happened in Maasailand is true of other pastoral areas in Tanzania:

> *The accomplishments in all these fields were meagre, according to evaluation reports of the various projects. ... The effort to change Maasai ways—both economic and social—has been massive in the sense that a large number of projects have been attempted, but it has been minimal in the sense that none of these projects—World Bank, USAID, and the country governments—has effectively incorporated the Maasai themselves into the planning and execution.* (ibid)

Many different donors are engaged in development projects in pastoral areas. Some are dealing with livestock development, some with agriculture, while others are assisting with social services such as schools and healthcare facilities. The provision of social services has made pastoral areas attractive to other occupational groups which had previously considered them to be unsuitable. Therefore the effect of these projects on tenure has been the withdrawal of large amounts of land from pastoralism.

MAIN AREAS OF CONFLICT

In some localities pastoralists are diversifying their activities, not as a way of abandoning pastoralism but as a faster way to acquire livestock and become independent. Agriculture and trade are the major alternative activities for pastoralists, especially young men. Those pastoralist children who go to school are taught about issues other than pastoralism. Those who manage to get administrative jobs become proficient on business matters. Yet the children who do go beyond primary school find it difficult to cope with pastoralism, because their education is geared towards agricultural production rather than livestock keeping. Therefore, when they return to their villages they prove less knowledgeable in this area than those who remained behind. This perpetuates inequality of opportunity.

The increase in agriculture means that pastures are destroyed and the seasonal herd movements are made more difficult. It also puts more land under the direct control of individuals and particular groups, and is thus a privatisation of pastoral resources. Once cultivated, an area becomes an exclusively owned resource outside the traditional economic and ideological system. Investments in agriculture and permanent buildings restrict the pastoralists to specific localities. Consequently, they are unable to make full use of the ecological diversity offered by their territories, and are forced into the overutilisation of particular areas.

Pastoralists in Tanzania are increasingly becoming a party to this, either in response to destitution or in pursuit of profitability. Among the Maasai, for example, several people have acquired hundreds of acres of pasture land for commercial farming. As these people increase in number so will the internal pressure on pastoral land. As agriculture is regarded by the government as one of the positive developments among pastoralists, this process is likely to continue at the expense of pastoralism unless specific areas are set aside exclusively for livestock keeping.

So far, private investment in the improvement of pasture land is still minimal. For example, water installations are still provided by the government. Although this 'free water' (Oba 1987) does not encourage communal responsibility for maintenance, it does help to maintain the communal character of the land. All dams and boreholes installed by the government (sometimes with contributions from the pastoralists) are communally accessible. Rich and poor alike can use the resources. Should an elite group intent on establishing its own exclusive water resources emerge, then another form of privatising pastoral land will develop. This occurred in Botswana when private wells and boreholes were established in communal grazing areas, and is said to have become a major factor in influencing political power and social differentiation (Gulbrandson 1987, Mazonde 1987).

To prevent a similar occurrence in Tanzania, the improvement of pastoral land should continue to be communal, possibly through the villages which are developing into stable social as well as residential units. If development funds are directed to a few so-called progressive individuals or groups, this will promote inequality.

Although falling per capita livestock ratios do not necessarily reflect overall poverty, they do indicate an increasing dependence on a non-pastoral diet. Domestic groups with less than the minimum subsistence livestock holdings have to engage in agriculture in which most of the work is undertaken by women. Because pastoral fields are 'cultivated' by men, the trend towards agriculture puts an increasing strain on pastoral women—as witnessed in Monduli Juu, Maasailand, where women are increasingly valued as a source of agricultural labour. Men in Monduli Juu are becoming more involved with agriculture by marrying more wives from communities familiar with agriculture, such as the

Waarusha. Though the wives do most of the manual tasks relating to agriculture, they do not seem to control the products. While the extent to which this happens is unknown, this trend is likely to continue as more land is given over to agriculture (Ndagala 1990b).

On the up side, some of the poor households which would otherwise continue to depend on the rich are now able to provide for themselves through agriculture. Moreover, because agricultural land can be allocated to whoever needs it irrespective of gender, more women can now secure their own means of production as individuals and not as wives, daughters or sisters. The seemingly widening gap between the rich and the poor pastoral households could therefore be bridged by agriculture.

ENVIRONMENTAL IMPLICATIONS

The sedentarisation of the pastoralists and the expansion of agriculture have placed many demands on the rangelands. The construction of permanent buildings, for instance, has raised the demand for wooden poles. The increasing dependence on agricultural foods has increased the rate of cooking which, in turn, has stepped up the demand for fuelwood. This demand is quickly decimating the wood cover in Maasailand, particularly near settlements. Women, who are responsible for fuel collection, now have to travel longer distances. Writing about the Garole Orma, Ensminger (1984) estimated that sedentary women use 60 per cent more firewood than nomads and that:

> ...*this, together with the increased distance walked in procurement, increases gathering time by an estimated 1300 per cent. Time spent fetching water also increases due both to greater distances travelled and the larger quantities demanded by the diet.*

The same may be said in respect of pastoral women in Tanzania. Women now have heavier workloads as a result of the change from a pastoral to an agricultural diet.

Another detrimental effect is the loss of tree cover in many pastoral areas. In places where the trees are of the traditional drought resistant type, which take many decades to grow to harvesting size, deforestation will lead to long term damage.

Non-pastoral uses of land, such as agriculture, wildlife conservation, charcoal and fuelwood accelerate environmental degradation. The clearing of land destroys plant and animal species, some of which, apart from their nutritional and therapeutic importance, perform important ecological functions. Largescale farming on rangelands, as is practised on Barabaig territory on the Basuto plains, is not only destroying valuable pastures, but is also exposing soil to erosion by wind and rain. The environmental effects of wheat farming on the Basuto plains are documented by Lane (1990a, 1990b, 1996). One of these effects is the silting up of Lake Basuto, which is an important permanent source of water for livestock and salt for the Barabaig and their livestock. The vulnerability of the pastoral societies is heightened by the fact that the dry season and reserve pastures, which are those usually lost through alienation, happen to be the best land.

Wildlife conservation in the form of national parks and game reserves features very strongly in and around pastoral areas, especially in northern Tanzania. Living amidst or near such an abundance of game has many implications for the pastoralists. Grazing livestock in the game areas is not allowed, but pastoral land is not protected from encroachment by wild animals. Marauding animals are a permanent threat to human life and to herds, and they compete for available grass and water, particularly during the dry season. Dangerous diseases such as malignant catarrh, common among wildebeests, are spread to livestock by game. Yet these wild animals have legally defined long-term security in parks and reserves, whereas pastoralists do not. This point is discussed in great detail by Parkipuny (1991) in respect of the Ngorongoro Conservation Area (NCA). He observes that to this day, the Ngorongoro Maasai have no effective voice in the Ngorongoro Conservation Area Authority. Moreover, he points out that the Administration of Ngorongoro has endeavoured to extinguish the legal rights of occupation of the people in the NCA (1991). This is a situation which needs policy review and legal attention.

CASE MATERIAL

There is very little published case material on the subject of pastoral land tenure in Tanzania. What is available in the official correspondence and reports is general in nature and advocates what Lane (1990a) calls the 'old orthodoxy'. In other words, it singles out pastoral land tenure as the problem. However, detailed case studies of the problems associated with land tenure are beginning to appeal to researchers, possibly as a response to the growing crisis in Africa's arid and semi-arid lands (Hjort af Ornas and Mohamed Salih 1989, Bovin and Manger 1990). Land tenure and related problems among the Barabaig are discussed by Kjaerby (1979, 1980, 1989) and, most recently, by Lane (1996). While Kjaerby looks at the change in land use among the Barabaig in general, Lane focuses on Hanang District. He describes in detail the traditional pastoral land use practices of the Barabaig and discusses how these have been affected by the massive land alienation by the state for wheat farming.

The problems of land tenure among the pastoral Maasai have been covered extensively. Parkipuny (1975, 1977, 1980, 1983, 1991) discusses the marginalisation of the pastoral Maasai by the state and criticises government policies towards livestock development in general, and pastoralism in particular. Arhem (1985a, 1985b) examines the land question in Maasailand, especially in the Ngorongoro Conservation Area, and the problems faced by pastoralists. He also looks at the changes taking place in the subsistence strategies of the Ngorongoro Maasai and their implications for pastoralism. The author's work (Ndagala 1990a, 1990c) takes up the question of land tenure and land use practices among the pastoral Maasai of Monduli District. The author has also looked at the problems of competition for land among the Ilparakoyo of Bagamoyo District (Ndagala 1974, 1986, 1991a). In both cases it was found that the continuous shrinkage of pastoral land was caused by the insecurity of pastoral land tenure in an area where land is needed by many different groups.

Data on the volume and value of pastoral production and the contribution this makes to Tanzania's economy are not readily available, and the few studies that have been done are mainly based on estimates. About 90 per cent of the total contribution of the livestock sector comes from the herd held by pastoral and agro-pastoral people (Tanzania Government 1982, Mustafa 1989). In 1977, Tanzania was estimated to have produced 340 million litres of milk of which 93 per cent came from the peasant smallholders. Arusha Region is the largest milk producer in the country with most milk coming from the pastoralists (Raikes 1981). If all the products of the pastoral sector were recorded the contribution of pastoralism to the economy would be found to be significant and worthy of serious consideration. Nevertheless, more detailed and reliable information is still needed before we can establish the exact contribution of pastoralism to GDP.

PROCEDURES FOR CONFLICT RESOLUTION

All pastoral groups have been in some form of conflict over land. Compared with the 1970s (Ndagala 1974, 1991b), there was a great deal more in the 1980s. For example, in 1985 a number of Datoga were occupied by agricultural communities in Iramba, Igunga, and Shinyanga Rural Districts (Ndagala 1991b). A special commission was formed by the ruling party, Chama Cha Mapinduzi, to look into the root causes of the conflict. The commission was formed after 48 people were killed at Mwamalole in Shinyanga Region (Ndagala 1991b). The commission visited nine districts and conducted interviews with members of the public and held meetings with party and government functionaries. Many officials were found to be prejudiced against the pastoralists, especially in those areas where the pastoralists constituted a minority. Beidelman's observation in the 1960s on the relations between the Ilparakuyo and their neighbours is still valid.

He points out that:

> *In general the present political system gives the Baraguyu few hopes of fair settlement in such inter-tribal disputes. Instead, they find themselves considered offensive intruders, even in areas such as parts of Kaguru, Gogo and Guu, which they have inhabited for over a century.* (1960)

In order to protect their land from further encroachment, pastoralists need new means of arbitration and negotiation. They need to learn how to deal with courts and other national institutions. They have to adjust to new demands, confront new threats and struggle to retain their resources. The Legal Aid Committee of the University of Dar es Salaam has on several occasions represented pastoralists in the law courts (Shivji and Tenga 1985). This aid should be stepped up now that the policy of economic liberalisation has encouraged commercial farmers to increase their holdings. In the meantime indigenous non-governmental organisations (NGOs) are being established by educated pastoralists in some areas to deal with land-related problems. One such NGO is Korongoro Integrated People Oriented Conservation (KIPOC), formed by the pastoral community in Ngorongoro District of Arusha Region. It wants to explore alternatives to land alienation, based on the compatibility of pastoral production and the conservation of natural resources. KIPOC hopes to help pastoral communities move from a subsistence economy to sustainable livestock and wildlife systems. It is hoped that these locally constituted NGOs will help pastoralists effectively to overcome the present problems. Observes Helland (1990):

> *The present situation of the pastoralists must be seen in a national political and economic context and their future will among other things depend on how well their problems and interests are articulated and understood in this environment of competition and conflict.*

The land problems of pastoralists in Tanzania was articulated through the work of the Presidential Commission of Enquiry into Land Matters. The commission, formed on 3 January 1991, had the following terms of reference:

1. to hear disputes from the general public concerning land and plots in the rural areas and urban centres, and to make recommendations for solutions;
2. to identify the basic causes of land disputes and to propose ways of settling them;
3. to review matters of policy and laws currently in force concerning allocations of land, land tenure, land use and land development, and recommend changes

wherever necessary;
4. to analyse the functions, jurisdiction and organisational structures of institutions involved in land matters, and to recommend the clear demarcation of their areas of responsibility;
5. to look into any other matters and issues connected with land which the commission deems fit for investigation.

With this commission the government seems to be committed to finding long term solutions to the different land problems in the country.

The commission completed its work in November 1992 when it submitted its report to the president. The report provides a most comprehensive statement on matters of land in Tanzania (Tanzania Government 1994). In the formulation of a new national land policy the government has considered the report. However, the commission's recommendation that radical title to land be divested from the president has not been accepted. Nevertheless, the new land policy supported by a complimentary basic land law will provide a revised and hopefully more effective framework for the administration of land.

CONCLUSIONS AND RECOMMENDATIONS

Pastoralists no longer have full control over the land they use. Since the state transformed pastoral land from its customary communal character into public land, most newcomers into pastoral areas no longer seek permission from the local councils of elders to use resources. Instead, the newcomers act on the strength of documents issued at the administrative headquarters at district, regional and national levels. Large tracts of land are continually being alienated and put on long-term lease for largescale farming. Moreover, population pressure in the agricultural areas continues to push landless cultivators onto pastoral land. As these alienations focus on areas of high potential (areas with, among other things, relatively higher rainfall) pastoralists are constantly being forced into poor areas.

Unless the situation is rectified, the conflicts over land will intensify. The problem is complex and needs to be handled sensitively. It is very easy for the support of customary rights in land to be misunderstood as tribalism. Should the Maasai or the Barabaig, for instance, have rights in a village or district and be refused access to land or other resources outside the area? If not, why should people from other areas be refused occupancy in areas predominantly occupied by these pastoral communities? In order to protect pastoral rights in land without compromising people's individual rights to live in areas of their choice, emphasis must be placed on the production systems and not ethnicity. This is an important point because already many pastoralists, like agriculturists, are living in places where they have not been customarily resident.

It is only when there is conflict that the inadequacy of customary land rights is realised. Sometimes this realisation comes too late. Given the ambiguity of customary land tenure due to the ethnic diversity in many areas, it is difficult to agree on which customs to use in resolving disputes. It is clear that this land will be alienated unless its occupancy is secured through law. Pastoralists should gain a commitment from the government that there will be no intervention in their use of land.

In the short term the following measures are proposed:

1. Further alienation of land in pastoral areas should be suspended immediately to allow the registered villages to determine their real land needs. This suspension should remain in force until all villages are surveyed, their boundaries demarcated, and their land protected by long-term rights of occupancy. This provision of legal rights in land will take a long time if it continues at the present pace. It was reported by the Minister for Lands, Housing and Urban Development on 7 July 1991 in Tanga that in one year only 146 rights of occupancy had been granted. And between June 1986 and July 1991 only 1293 villages out of the 8000 villages in the country were provided with rights of occupancy. Unless alienation is suspended some villages will hardly have any land by the time they are due for survey.
2. The villages should be helped to produce land use plans for pastoralism, along with agriculture, and other production strategies.
3. The villages should enact by-laws that will ensure that their members protect the environment.

Implementation needs the cooperation of many people. Researchers, legal experts, planners and NGOs all will have to help the pastoralists take control of their resources in a sustainable way.

The registration of titles for the different villages may cause problems. Some resources, such as water and dry season pastures, which are shared by people across a wide area, may end up being part of a particular village to the exclusion of others. In such cases members of the other villages who used the pasture under a common property regime will have to negotiate ways of utilising the resources in question. This is already common practice in many pastoral areas. Eventually, each village will evolve its own grazing scheme and share an understanding with neighbouring villages on how to manage and utilise these shared resources. Registration of the villages and the issuing of titles will change pastoral land from being public into legal common property of the village communities.

Longterm pastoral development plans will depend on the availability of sufficient land and on detailed information on its ecology and the uses to which this land can be put. There is need for more data on:

- changing property rights within pastoral groups;
- decision-making processes and the execution of power at the local level;
- production diversification and resource re-allocation.

With this information it will be easier to see pastoralism in the context of the whole of society, rather than in isolation.

There is a greater need now than ever before for social scientists to work together with veterinarians, agronomists, range managers, agrologists and so on, and to make clear their findings to decision-makers.

> *Many of the problems of communication between decision-makers and social scientists stem from the fact that most social scientists do not know the language of development and do not do their homework on ethnography of development. They are not sympathetic to the kinds of problems that planners face.* (Horowitz 1981)

Finally, nothing is likely to succeed unless pastoralists are themselves fully involved in identifying and solving their problems of resource availability and management.

---

[1] The views expressed in this chapter are those of the author and not the Ministry of Education and Culture.

8

UGANDA

W Kisamba-Mugerwa
Makerere Institute of Social Research, Kampala

Map 25 Uganda

INTRODUCTION

The traditional pastoral areas in Uganda (Map 25) are experiencing degradation of their natural resources. Management crises related to rangeland tenure and resource use produce a host of environmental, social, economic and political problems. Pastoral areas have been encroached on by non-pastoral development, and this has put pressure on the remaining grazing areas. These have made pastoralists vulnerable to resource degradation in general and overgrazing in particular, caused a decline in production and heightened conflicts over use of resources, and provoked social and political instability. Overall, the welfare of traditional pastoralists in Uganda, comprising the Bahima and Basongoro in the southwest and the Karimojong in the northeast, is poor relative to other Ugandans in terms of social service facilities.

In Uganda, cattle provide a major source of food and means of livelihood. Projections for 1995 drawn from the 1991 Livestock Census indicate the considerable size of the national herd of five million cattle, nearly half a million goats, and over a million sheep (Government of Uganda 1992). After a decline in livestock numbers due to civil unrest, general insecurity and cattle rustling in the north and northeast of the country, there has been a resurgence of growth in the contribution the livestock sub-sector makes to GDP, which is currently increasing at 4.9 per cent per annum (Government of Uganda 1993a).

Management of rangeland resources for sustainable development remains one of the major challenges facing policy analysts and development agencies. Both colonial and post-independence governments have invested funds in isolated components of pastoral development but have failed to achieve sustainable resource use. Instead, they have tended to disrupt traditional pastoral rangeland tenure by introducing resource management systems which threaten food security. Governments have also restricted social benefits to pastoralists and minimised their involvement in local planning.

The arid and semi-arid areas of Uganda include the traditional grazing areas in what is referred to as the cattle corridor (Map 26). These areas stretch from the southern Uganda border with northern Tanzania through eastern Mbarara, eastern Masaka, covering the areas of Ngoma, Nakansongola County in Luwero District, Baale County in the northern part of Mukono District, and the eastern parts of Hoima and Masindi districts. The area also extends from the northern parts of Kamuli District to Lake Kyoga and beyond, to Soroti, Moroto, and Kotido districts in northeastern Uganda.

POLICY AND LEGAL CONTEXT

There is no one land tenure pattern throughout Uganda. Customary land tenure, practised in the pre-colonial period, varied from one ethnic group to another. However, many customary patterns, particularly in pastoral areas, were based on forms of communal ownership.

No single land tenure system was established after the 1900 Uganda Agreement, which allocated land to the king, chiefs and notables in Buganda

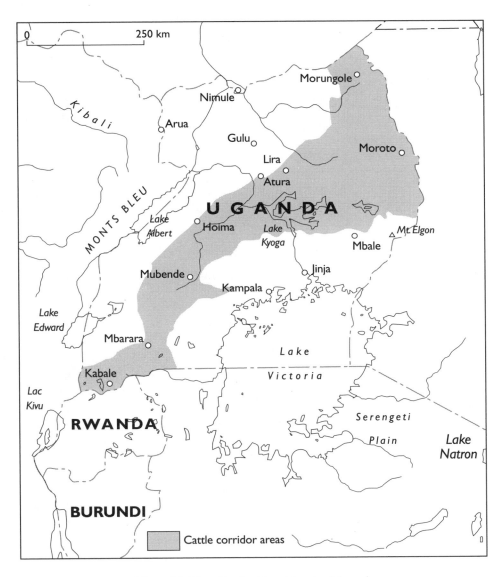

Map 26 The Cattle Corridor in Uganda

(area north of Kampala occupied by the Baganda people). Agreements in Toro in 1900 and Ankole in 1901 were similarly inconclusive. Although Uganda was a protectorate rather than a colony or territory, the government's policies toward the indigenous tenure system were far from indirect. The introduction of *mailo* tenure in Buganda in 1900 (a system akin to freehold land tenure) was accompanied by the introduction of freehold tenure in Toro and Ankole, locally referred to as 'native freehold'.

The Crown Lands Ordinance of 1903 gave the British colonial authorities power to alienate land in freehold. Though few freeholds were introduced under this ordinance, together with leaseholds introduced on crown land, the legislation radically transformed the customary tenure system (Mugerwa 1973, Richards et al 1973, West 1964, 1972, Beatie 1971). The colonial authorities believed that customary tenure systems were prone to insecurity. It was thought that under the system clan and community were unable to invest in the land as a corporate body, while individuals with initiative lacked incentives to make improvements. The main concern of policy-makers during the colonial period was centred on how to make Uganda self-reliant in terms of administrative costs, while ensuring a supply of raw materials for industry in Britain. The aim was, therefore, to replace extensive cattle keeping with cultivation agriculture.

'Native freeholds' were set up by the Ankole and Toro agreements of 1900 and 1901 (Morris and Reed 1966). Each of these agreements carried a land settlement provision which set out a distribution scheme for the kingdoms. Though the freeholds were restricted, they initiated a policy which transformed pastoral rangeland tenure by encouraging pastoralists to settle.

The prime concern throughout East Africa was to make the administration self-financing. Agricultural development through cash cropping was introduced not only to provide raw materials for industries in Britain but also to form a tax base to finance local government. The report of the East African Royal Commission in 1953, produced in 1955, made the following recommendations:

- land tenure policy should seek the privatisation of land ownership;
- transactions of land should enable easier access for economic use;
- land tenure should not be allowed to develop spontaneously; government should guide its development to meet the needs of a modern economy;
- existing property rights in land should be maintained and customary land rights must be ascertained and accommodated before exclusive individual rights are sanctioned;
- land registration should not promote subdivision and fragmentation;
- land tenure reform should accommodate local circumstances and be pursued only with local support.

These recommendations were officially accepted by the Uganda government, in particular the recommendation that land tenure should be based on individualised freehold title. Though these recommendations were subject to veto by each local administration, they effectively enhanced the process of individualisation. Some pilot schemes were undertaken in Buyanja sub-county in Rukungiri District; in Kagango and Shuku sub-counties in Bushenyi District; and in Bugishu which is now Mbale District. Notably, the pilot schemes sparked off sporadic surveys of individual parcels of land (Kisamba-Mugerwa et al 1989). Land which had not been registered either as *mailo* land or as another type of freehold was classified as public land under the colonial administration.

After independence, provisions for protecting customary land rights were provided by the Public Land Act of 1969. A person could occupy, in customary tenure, any rural land not under leasehold or freehold. The controlling authority could only grant a freehold or leasehold on any land occupied by customary tenure with the consent of the customary holder.

A fundamental legal change was ushered in by the 1975 Land Reform Decree (Khiddu-Makubuya 1991). This declared all land in Uganda to be public land vested in the Uganda Land Commission. It abolished freehold interests except where such interests were vested in the commission. As a result, all freehold land, including *mailo* ownership which existed before the decree, was converted into leasehold.

For pastoralists the conversion of freehold into leasehold made no difference. Individual acquisition of land under the Land Reform Decree continued. Moreover, the decree lifted the basic legal protection which had been enjoyed by customary tenants on public land when it allowed the controlling land authority to alienate any public land occupied by customary tenants, without their consent. In this way the decree left the customary tenant on public land in a very precarious position: no person could occupy public land by customary tenure except with the written permission of the prescribed authority (Section 5(i)).

Clearly the thrust behind government policy was to introduce private ownership of land deemed suitable for production. The degree of individualisation, however, has varied from one ethnic grouping to another. Among the pastoral Bahima in Mbarara it is almost certain that communal grazing is being phased out, while in Karamoja private ownership of land is being resisted. A branch of the Department of Lands and Survey was opened in Moroto in early 1991, but the officials then left and the office no longer functions. A district land committee has, however, been formed. It deals almost exclusively with allocation of plots in the urban areas, and has no business in rural areas except where NGOs have acquired land for their projects.

Due to the difficulties being experienced with dependency on livestock in Uganda, especially among the Karimojong, cultivation is becoming widespread. During the short rainy season plots of sorghum can be seen near their homesteads. These are prepared and looked after by women. With cultivation comes a move towards enclosure, possibly followed by survey and registration of title.

An examination of the laws that have governed the administration of land in Uganda reveals how communal land tenure has been gradually transformed by development policies for the privatisation of property and the intrusion of the market economy.

The maximum interest possible under the Land Reform Decree is a leasehold title. The former *mailo* land and freeholds were supposedly converted into 99-year leases in the case of individuals, and 199 years in respect of public bodies. Customary tenure was deemed to continue on public land subject to termination on terms approved by the minister. Though in law all land is public land, in practice the *mailo* and freehold registered certificates of title are recognised and therefore the decree has never been fully implemented as the *mailo*/freehold titles concurrently run together with leasehold titles. From time to time the government has intervened to prevent the implementation of the law. Several cases have been taken by the government to stop the evictions of customary tenants (Nsibambi 1989). This has remained a source of controversy among lawyers, administrators and politicians.

Because of the difficulties in administering the Land Reform Decree, and its counterproductive effects on agricultural development due to the increase in social tension it has created, a study of land tenure in Uganda was initiated at the Makerere Institute of Social Research (MISR), at Makerere University. This covered land tenure, settlement in national reserves, and the effects of land-titling on agricultural development. After several workshops a technical committee was chosen to review the proposals laid down by MISR.

The proposed land law reform policy recommends private land ownership, in the form of freehold, throughout the country. Although it recognises customary rights to common land it recommends formal registration of the communal grazing land in the names of an organised group of rangeland users.

Traditional pastoral land tenure has not been given adequate attention in Uganda. The focus on land tenure policies has been centred on farm land that is clearly demarcated and where improvements are easily discernible. The movement of pastoralists and their herds in search of water and pasture over a common range is thought to be unproductive. Investment in improvements, such as burning grass to rejuvenate pastures, is seen as a destructive practice.

The negative stereotyping of pastoralists as 'nomads' was reflected in the appointment of a 'minister of state in charge of ranch restructuring, water development, and anti-nomadism' in the National Resistance Movement cabinet reshuffle of 18 November 1994. Such preconceptions tend to lead to the adoption of inappropriate policies and development approaches that fail to deliver adequate social services and technical support to pastoralists relative to other groups.

It is thus clear that Uganda lacks a coherent national land use policy. Independent disciplines have focused too much on land as a commercial resource. For example, livestock policies are biased towards ranch development for commercial purposes. In addition, conservation strategies have focused on the preservation of wildlife without taking into account the needs of other forms of land use. In the past, pastoral areas have been the focus of water improvement and livestock health facilities. However, broader issues of pastoral land use have generally been ignored. An exception has been the Ranch Restructuring Board, commissioned by the government, that set out to accommodate landless pastoralists on ranches which are considered large and undeveloped in Ankole/Masaka ranching schemes.

Another example of action is the Karamoja Development Agency, which was established by the government in 1987 to supervise rapid economic and social development in the area. This agency is to ensure that the Karimojong people acquire the necessary skills to enable them to bring about this development. The agency is specifically charged with providing sufficient water for agriculture and livestock husbandry, and with a view to settling them.

TRENDS IN PASTORAL DEVELOPMENT AND WELFARE

Pastoral areas in Uganda are found in arid and semi-arid savanna grasslands punctuated with thickets and an isolated chain of small lakes, particularly in the Ankole region. The terrain becomes more arid in the northeast area of Karamoja. These areas experience unreliable rainfall with a long dry spell from October to March. The mean annual rainfall varies from 500mm to about 1000mm with a high degree of fluctuation between years and locations. Average temperatures range from 18°C to 20°C with a maximum of 28°C and 30°C. These areas are sparsely populated by the Bahima of Ankole, and the Jie, Labwor, Matheniko and Upe of Karamoja.

In the traditional pattern of pastoralism, wildlife has coexisted alongside pastoralism. However, in 1894 the Commissioner of Uganda noted the decline in wildlife and proposed measures to protect certain species. The creation of national parks was a result of recommendations made by Ken Beaton, the pioneer warden of Nairobi National Park, on the invitation of the then governor of Uganda, Sir Andrew Cohen, in 1951. The National Park Act was passed in 1952

(UNEP 1988). The effect of this legislation was to protect wildlife at the expense of pastoralists. Uganda now has numerous national parks and game reserves, many of which cover much of the cattle corridor. The boundaries of the game reserves and national parks, however, have no ecological basis and do not effectively confine wildlife. Reports indicate that wild animals, particularly ungulates, graze extensively outside conservation areas.

The whole of Karamoja Region is a controlled hunting area. It is also the location of the Kidepo Valley National Park, Matheniko Game Reserve, Bokoro Corridor Game Reserve, and Pian-Upe Game Reserve. The game reserves in Karamoja cover 6908km^2 alone. Lake Mburo National Park in the southwest of Uganda is of particular concern to pastoralists. This is because when it was converted from a game reserve into a national park, all human activities other than those connected with the management or utilisation of wildlife resources were strictly prohibited, despite the fact that the lake provided one of the few permanent sources of water for the local population. It is important to note, however, that in game reserves in Karamoja, pastoralism is, to a certain extent, permitted and cattle still coexist alongside wildlife.

Government agencies, NGOs and other donor agencies have supported development projects in pastoral areas. USAID took a leading role in the eradication of tsetse flies and the subsequent establishment of the government-sponsored ranching schemes throughout the Ankole area. It was agreed between the government of Uganda and USAID, as a condition of the 'Ankole Tsetse Fly Eradication and Control Scheme', that priority would be given to land for cattle raising and that a land tenure system permitting rational land use would be established in the area. The same was reflected in an agreement signed in March 1992 between the government and USAID, in which the initial strategy was to bring into production large tsetse fly infested, sparsely populated and low rainfall areas.

The government's main objective in setting up ranching schemes was improved animal husbandry practices. Several commercial ranches with improved beef cattle were established in the areas cleared of tsetse fly. By 1968 plans for ranch development were undertaken and a field survey of roads, valley water tanks and construction work had been completed. Following the same pattern, the then local government of the Buganda Kingdom in the central region developed ranching schemes in Buluri, north of Luwero District, and in the Singo area in Mpigi District.

These developments were in the interest of commercial livestock production but not necessarily the pastoralists who were displaced. As Lane observes (1996) with the displacement of Barabaig pastoralists in Tanzania, the decisions about transforming pasture into farm land should have included an economic analysis assessing costs of withdrawing land from the pastoral system. USAID supported the eradication of the tsetse fly and the subsequent establishment of ranching schemes on the understanding that the Uganda government would not implement the scheme beyond the guidelines laid down in the original technical report. The Gregory Report, as the USAID technical report came to be known, emphasised that ranchers should actually reside on the ranches. This would have eliminated absentee landlords and accommodated more of the indigenous pastoralists who were using the rangeland before the ranching scheme. This remains a source of controversy in the development of government-sponsored ranches in Uganda. Displaced pastoralists have, of late, tried to assert their rights to ranchland, causing serious social tension. This has resulted in the commissioning of a committee of enquiry into the ranching schemes and the subsequent formation of the Ranch Restructuring Board.

Many NGOs and government agencies have a substantial collection of unpublished data on pastoralism which, if synthesised, would shed light on pastoralism in Uganda. The process of development in the pastoral areas in Uganda has been diverse. In the southwestern part of Uganda, particularly in Ankole among the Bahima, the emphasis has been on developing the beef industry and the associated industries of leather tanning, animal feed, ghee and butter production. A meat canning factory was built in Soroti District for this purpose. In northeast Karamoja the colonial administration was preoccupied with maintaining peace among the Karimojong (Baker 1967, Cisternino 1979). This still remains a priority.

The lack of coherent development in Karamoja has made the government encourage NGOs to support development there. In 1987, the KDA was launched by the government to coordinate and direct development activities in the region, which since the 1950s have been in the form of construction of valley water tanks, dams, bore holes, smallscale irrigation schemes, and agriculture and technical education.

These developments have, in most cases, worked to weaken pastoral land rights. In some cases, as in the ranching schemes, developments have physically displaced pastoralists. This is particularly true in Nyabushozi County in Mbarara District. Game reserves and national parks have excluded pastoralists. Other development projects, especially those providing water, have disrupted land management by concentrating people and cattle, resulting in the subsequent overgrazing of pasture. This is particularly evident in Karamoja where herds are assembled around water points in the dry season.

MAIN AREAS OF ALIENATION AND CONFLICT

Pastoral areas have always been seen as areas of low economic development. The Commission of Inquiry into Government Ranching Schemes noted that the inception of commercial ranching, which dates back to the late 1950s, had the specific objective of meeting national demands for economic development. The strategy was aimed at bringing into production country infested with tsetse fly in sparsely populated and low rainfall areas. It was emphasised that the agricultural potential of vast areas of Uganda was being underutilised. The government cleared tsetse flies, demarcated the area into blocks of five square miles each, and allocated ranches to promote the national beef industry.

During the period of ranch demarcation, pastoralists were unaware of the opportunities to register titles to land, or of the adverse effects of establishing ranches on their land. Even the selection criteria for ranch allocation were unclear to them. The procedures for application were too elaborate for an illiterate population to follow easily. Advertisements in newspapers, published in English, remained largely inaccessible to them, and the terms and conditions of occupancy of the ranches were also not made clear by the administration.

The outcome of this was that large numbers of pastoral people were rendered landless, while retaining large numbers of livestock. Displaced pastoralists have had to settle with their herds as 'squatters' on ranches and in national parks, forest reserves and other private land (Pulkol 1991). The result has been growing conflict between displaced pastoralists and commercial ranchers.

Land shortage, due to population increase and encroachment by farmers in pastoral areas, and land taken for the conservation of wildlife, has driven some pastoralists to adopt agro-pastoralism. This has brought about greater social and economic inequalities among pastoral societies, as some pastoralists are being transformed from independent rural producers into dependent herders of other people's livestock, a situation already well documented in other places in West and East Africa (Bennett et al 1986).

Another area of stress arises from the immigration of farmers into pastoral areas. The population of Rushenyi County, particularly Ngoma and Rubaare sub-counties, in Bushenyi District, is made up of over 60 per cent of people from Kabale and Rukungiri districts. Many of them are said to be of Rwandese origin. In other sub-counties, cultivation is mainly undertaken by immigrants. For people from heavily populated areas where land shortage is acute, as is the case among the Bakiga from Kabale District, a parcel of land of about ten acres doubtless seems a large piece of land for cultivation. With a strong background in cultivation, they are eager to settle on the most fertile land in pastoral areas, and to facilitate this they offer attractive prices for land. This induces pastoralists to leave for other, more arid, areas in the hope of having access to larger areas of land.

The counties of Nyabushozi and Kazo in Mbarara District are increasingly being settled by people from Rukungiri and Bushenyi districts. In some of these areas, especially between the Rubaare and Ntungamo trading centres along the Mbarara—Kabare road, indigenous people have formed cooperatives to manage their grazing areas communally. These are, however, disintegrating due to the individualisation of landholdings.

Another area of resource use conflict is the refugee settlement scheme in Bukanga and Isingiro counties in Mbarara District. The area has two settlement camps, one at Nakivale in Bukanga County stretching over 84 square miles, and another at Orukinga, partly in Bukanga and partly in Isingiro counties, covering about 13 square miles. By April 1991 Nakivale had a total population of 14,000 refugees with about 35,000 head of cattle, while in Orukinga there were 4949 refugees with about 25,000 head of cattle.

Similar refugee settlement camps are found in Kyaka County of Kabarole District. In all cases open-access grazing is practised. This is due to the breakdown of traditional controls in the settlement areas. The areas are, therefore, susceptible to overgrazing, particularly in Nakivale and Orukinga settlements where erosion is common. Indigenous people blame this overgrazing on refugees, and question giving land to the refugees who were once thought to be only temporary residents.

The creation of national parks and game reserves has also provoked conflict. Lake Mburo National Park provides a good case study in this context. At present the park is managed according to a 1989 development plan, based on the National Park Act. The act does not allow any form of pastoral utilisation of land (a park census in 1991 revealed a resident population of 1672 people with 16,517 cattle, 150 goats, 7 sheep and 7 dogs). The planned eviction of 'squatters', and the denial of access to local herders, has created negative attitudes and antipathy to wildlife among the local population. Resolution of this conflict depends on the achievement of more integrated development, and the longterm conservation of wildlife resources in the area.

Besides commercial ranching schemes the surveying and registration of private title to land is becoming increasingly common, especially among the Bahima in the Nyabushozi/Kazo area. A few individuals in Karamoja (mainly at Namaluwho), who leased parcels of land, have had this land overrun and their fences uprooted by the local pastoralists.

As yet local government, entrepreneurs, traditional leaders and even political parties have not featured in pastoral matters. In Uganda it is the central government and NGOs who have played a part in policy and development. Before conflicts can be resolved, all parties must become more involved and have the opportunity to influence development in areas and on issues that affect them.

ENVIRONMENTAL IMPLICATIONS

Various development projects and programmes undertaken in arid and semi-arid areas have affected pastoral resources. The degradation of pastoral resources, however, is alarming. Overgrazing, lack of vegetation and low livestock productivity have all affected the welfare of the people in these areas.

An examination of rangeland management in Uganda reveals significant environmental hazards. Pastoral areas, being marginal lands, require balanced land use and the pressures of an increasing population only serve to compound the problem. In its mild form, degradation of grazing land causes loss of the most nutritious forage species which are replaced by less nutritious and palatable species. This has occurred widely in Ankole and Karamoja districts.

The impact of settlements on areas traditionally used for grazing is usually profound. Rushenyi County in Bushenyi District has had an influx of immigrants from the more densely populated areas of Kabale, Kisoro and Rukungiri. On account of this intrusion the remaining grazing areas of Rubaare and Ngoma have become overgrazed. One effect is a reduction in the vegetation cover. In some cases the land is left bare of any vegetation, as has happened in areas surrounding the Lake Mburo National Park along the Rwizi river. Severe soil erosion has created gullies in some parts of Nyabushozi County in Mbarara District.

The cultivation of marginal land near to new settlements has contributed to degradation around the towns of Moroto and Kotido in Karamoja. The area experiences high temperatures during the rainy season and desiccating winds during dry spells. Drought manifests itself in crop failure for about four out of ten years. In these circumstances soil remains unprotected against wind. When rain falls it carries soil into dams and as a result most of the dams constructed in the 1960s, both in Ankole and Karamoja, are now silted up.

Pastoralists tend to move their herds away from the main homesteads during the dry season. Such transhumance is quite common especially in Mathniko, Jie and Upe counties in Karamoja, and to a certain extent in Nyabushozi and Kazo counties in Mbarara District.

Access to land for grazing livestock is uncontrolled in refugee settlements such as Nakivale, Oruchinga, Kyaka I and Kyaka II. Refugees are not allowed to claim any parcel of land for grazing purposes, but are allowed to have open access to the grazing land within camp boundaries. This arrangement has adversely affected natural resources. Due to the uncertainty of tenure, refugees in these settlements have never attempted to curb overgrazing.

In the Lake Mburo National Park, overgrazing is clearly visible around the lake and along the River Rwizi where areas have been heavily overgrazed due to the permanent source of water. It is estimated that one-eighth of the park has been overgrazed. Nomadic and semi-nomadic cattle keepers who pass through the park from the neighbouring sub-county of Gayaza in search of water are held responsible. The herders seem to be aware of the danger of overgrazing and for that reason they do not settle in any particular location. A similar situation is being experienced in Katonga Game Reserve.

In an effort to curb overgrazing the Lake Mburo National Park authorities have restricted settlement of 'squatters' to within 2km of the park boundary. Demarcation of the boundary has not taken place because local politicians and administrators have prevented it. The claim for a cattle corridor through the park was deemed unjustified by the park authorities, because ranchers had access to Lake Kakyera on the eastern side and the Rwizi river on the western side of the park. However, the Ranch Restructuring Board has facilitated the movement of cattle to Lake Mburo, Lake Kakyera, and to large dams north of the ranching scheme (*New Vision*, 13 July 1991).

Overgrazing on the government-sponsored ranches has been mainly brought about by squatters who have encroached on the ranches due to lack of grazing land elsewhere. The Board has observed that soil erosion caused by overgrazing is occurring in nearly all communally grazed areas outside the ranches.

Very little bush clearance takes place on both government and private ranches. Rotational grazing is not practised. In Nyabushozi and Kazo counties only on the ranch of the president, General Yoweri Museveni, have pasture improvement measures been taken. Dryland savanna is common throughout the area. In more fertile areas some wet-tree savanna species are found, but only on some private ranches is attention given to pasture improvements and bush clearance. Otherwise only low quality pastures are found.

Uganda lacks a systematic method for assessing the impact of policies and programmes on the potential for sustainable development. No one in Uganda has assessed the cost of policies which allow continued environmental degradation and depletion of natural resources. Without this assessment it is difficult for policy-makers to appreciate fully the value of environmental protection, or to have the means to make decisions which involve trade-offs between economics, the environment and social equity (Slade and Weitz 1991).

THE ROLE OF WOMEN

Traditionally women in pastoral areas hardly have any control over livestock, particularly cattle. They are more likely to own small stock, although in some cases individual cows might be owned. Being responsible for feeding the family women are assured access to livestock products, and this involves them in milking cows, preparing dairy products and caring for calves.

As traditional economies have become more commercialised, particularly in the southwest among the Bahima, women have begun to trade in livestock. Women have also taken up cultivation as a means of supplementing the pastoral economy, and increasing monetary demands have forced women to harvest fuelwood for cash. However, despite a changing role in society for pastoral women this is not always reflected in the affirmation of their rights to resources. This issue is a matter of concern when communal lands are privatised as women inevitably miss out on allocations and this leaves them vulnerable and marginalised.

Quam (1976) emphasises the motivations and incentives which lead groups to adopt certain actions. Baker (1975) notes that during the period of external administration, symptoms of the problem rather than the problem itself were treated. Water development, disease control, destocking—all became ends in themselves, unrelated to the overall needs of pastoral society.

Most of this work contains little quantitative data, particularly in respect to pastoral production and the contribution it makes to the national economy. With the exception of the scanty records of offtake found where cattle markets are established, the overall understanding of pastoral production is absent. Different development agencies, such as Oxfam and the Lutheran World Federation, have compiled substantial qualitative and quantitative information on various aspects of pastoralism. Now that conditions for research are gradually improving, comprehensive studies on pastoralism have been undertaken by both MISR and the Independent Centre for Basic Research in Kampala.

The approach to development planning in Uganda has for a long time been based on isolated projects and it is not uncommon to find duplication of activity. A more integrated approach to development planning, leading to a coherent national land use policy, is therefore urgently needed.

CASE MATERIAL

The security situation in Uganda since independence has made it very difficult to carry out research. Studies on pastoralism in Uganda have mainly been confined to anthropological or administrative aspects, and were mostly carried out during the colonial period. The main purpose of these studies has been to support the transformation of pastoralists into settled peasants through the introduction of ranching and private land ownership.

Other material has been written by missionaries who took useful notes during their long stay among pastoralists. Case material originating from unpublished project reports made by students in fulfilment of their first degree courses, and reports by professionals working with NGOs, also provide useful additional sources of data.

Very few studies, however, have specifically examined the problems associated with rangeland resource use. What has been published is mainly about Karamoja. The most celebrated work was by Dyson-Hudson (1958, 1962) which focused on the Karimojong. In addition the Bahima have been well documented by Karugire (1971), and traditional pastoral land use practices and the changes occurring among the Bahima have been analysed by Doornboss (1975). A case study by Doornboss and Lofchie (1971) analysed the controversy about the allocation of the government ranches in the Ankole ranching scheme.

PROCEDURES FOR CONFLICT RESOLUTION

The government approach to conflict over land use varies with each case. In the case of Lake Mburo National Park, the government reduced the park from 250 to 100 square miles to accommodate pastoralists and ranchers. Although by law no human activity is allowed in the national parks, in Lake Mburo National Park livestock grazing is still tolerated. In Karamoja region, grazing in game reserves is also permitted.

Following the report into government ranching schemes, the Ranch Restructuring Board scaled down ranches to three, two and one square mile/s. The board is also meant to resettle landless pastoralists and to provide a longterm policy regarding the management and development of natural resources.

Within the mechanism of the 1975 Land Reform Decree, the government has often intervened to reverse cases involving evictions. Doornboss (1975) concludes his paper by quoting president Idi Amin, in whose regime the notorious decree of 1975 was introduced. He said that landowners who had bought large areas of land in the past should not evict their tenants on short notice (*Uganda News*, No 4220, 1972). The government's intervention often makes administration of the decree difficult, although its purpose—to avert social tension—has some validity.

Pastoralists are not represented by any organisation which could protect their interests and facilitate their use of the legal system. The Land Decree does not recognise customary occupancy of land, and is biased against pastoralists who wish to graze their livestock on a rotational basis on common land. The only way open to pastoralists to seek redress is to ask the government to intervene on their behalf. Fortunately the current National Resistance Movement (NRM) government has some sympathy with pastoralists in solving their conflicts. This is exemplified by the manner in which the government responded to a dispute over land between Basongora pastoralists and Bakonjo farmers in Kasese District in which a cabinet committee conducted a study and reported on ways to avert the conflict (Government of Uganda 1993b).

In Karamoja, widespread cattle raiding has created widespread insecurity within the area and neighbouring districts. Because of its extent, the government has employed the army to intervene when ordinary administrative measures have failed. Recently, however, the government has enlisted the help of traditional leaders in Karimojong society to register privately owned guns as a means of controlling conflict.

RECOMMENDATIONS

1. Before any development project is introduced in a pastoral area the local community must be fully involved. The intervention must be designed to fit within the local framework, and be consistent with existing pastoral values.
2. The Parks Act and the Game Preservation Act need to be amended. This is one of the most pressing issues that needs investigation, and should lead to the formation of realistic, sustainable resource use programmes. Conservation projects designed for pastoral areas should not displace local communities because it is more difficult to find a suitable formula for fair compensation than it is with cultivators.
3. Specific studies that explore resource management and the nature of local resource use are required. It is important to find out to what extent pastoralists have adjusted to the changing socioeconomic conditions related to land tenure, especially within the market economy. For example, there is no data on how different property regimes affect pastoral resources in terms of range management, productivity, efficiency and equity, nor how conflict resolution mechanisms regarding rangeland tenure and use have evolved.
4. The position of women in the pastoral economy needs rigorous investigation and analysis. While women may own cattle they are generally overlooked in the allocation of land.
5. Pastoral systems are experiencing profound changes due to the development of the commercial economy. Since the state's interest in pastoralism is confined to productivity, the pastoralist's contribution to the national economy needs to be quantified.
6. The most striking observation of all is the degradation of resources found throughout all rangeland areas. The root cause of overgrazing under different property management strategies needs to be examined. The lack of a national land use policy has intensified resource use conflicts. Before these can be

resolved the environmental database will need to be strengthened. A case study approach to rangeland resource use and different management practices is a prerequisite to a successful integrated development strategy in Uganda.

Appendix I

Guideline

AN OVERVIEW OF THE PROBLEM OF PASTORAL LAND TENURE IN AFRICA

INTRODUCTION

The International Institute for Environment and Development and the United Nations Research Institute for Social Development wish to explore the issue of land tenure and its impact on resource management and environmental degradation in pastoral areas of Africa. Their interest is prompted by research findings that suggest changes taking place on the rangelands are undermining traditional pastoral livelihoods. While there may be many causes responsible for this, the purpose of this research is to identify the role played by land tenure in this context.

BACKGROUND

Pastoral people who rely on drylands for their living have always suffered recurring droughts and times of food shortage. However, despite scarcity of resources and vagaries of climate, it seems that people in the past were better able to survive. This was achieved in part by systems of common land tenure which enabled them to make the most use of scarce and scattered resources and to cope during periods of drought.

Indications are that conditions for traditional pastoralists have worsened considerably: ever increasing areas that were once communal pastures have been lost to pastoral production; irrigation schemes, smallscale farming and mechanised agriculture have withdrawn large tracts of the most productive land; food production per head and living standards have fallen; future incomes and welfare are further threatened by increased degradation of natural resources, while a growing conflict of interests is pitting pastoral communities against governments and other land users.

RESEARCH AIM

The aim of this research programme is to assess the problem of land tenure for pastoralists throughout Africa, and to identify how best to stem the decline and to secure sustainable livelihoods for pastoral peoples in the future.

The purpose of these studies will be to compile an overview of the problem so as to: guide future research; design alternative policy interventions; and support pastoral organisation.

TERMS OF REFERENCE

Each research study will describe the extent and nature of the problem of land tenure as it affects pastoralists within a specified area. Please indicate how your contribution (authors) could address some of the questions outlined below:

Policy and Legal Context

1. What are the explicit and implicit theoretical, ideological and political bases on which are formulated policies and laws relating to land tenure (and in particular pastoral land tenure)?
2. Are there misconceptions and prejudices about pastoralists' way of life and production? Is this reflected in drives for cultivation agriculture and export crops that exclude benefits for pastoralists?
3. Do state laws adequately support the concepts of communal land tenure and customary rights to land? What legal rights to land currently exist for different groups? To what extent is pastoral land tenure defined in a different way from farmland tenure?
4. What national policy instruments—livestock development policies, conservation strategies, land use plans—are relevant to pastoral land tenure? What are the trends in pastoral development and welfare?
5. Are developments on the rangelands a product of these policies? If not, how and why do they differ?
6. What aid donors are involved with development in pastoral areas?
7. What effects have development projects had on land management, use and tenure arrangements for pastoralists and other groups? Have pastoral land rights been strengthened or weakened by such projects?

APPENDIX I

Main Areas of Alienation and Conflict

8. How have changes within pastoral society brought about modifications to land tenure issues? To what extent are difficulties with management of pastoral resources and grazing lands due to changing power relations within their societies?
9. How have changes in land tenure affected pastoral women? Have they lost access to land or livestock and, if so, how has this affected their economic status? How is this reflected in women's position within the community?
10. In what ways do local government, entrepreneurs/traders, traditional leaders, political parties determine access to pastoral resources? Outline environmental impacts.
11. To what extent are pastoralists contributing to environmental degradation? If so, what form is it taking? How is this being revealed or measured?
12. Are non-pastoral uses of what were once pasture lands causing degradation? What form? How measured?
13. What factors are causing or augmenting environmental stress and how are pastoralists coping?

Case Material

14. Does published case material exist that examines the problems associated with pastoral land tenure?
15. What is known of traditional pastoral land use practices, and the changes that are occurring to these systems? Have these been documented, and if so where?
16. What data are there that quantify pastoral production and the contribution it makes to national economy?
17. Have economic and environmental costs of alternative uses of pasture land been recorded?

Conflict Resolution

18. What is the government reaction to conflicts over land involving pastoralists?
19. In the case of disputes, do pastoralists have the knowledge, ability and contacts to make good use of the existing legal system? Are they adequately represented in local and national decision-making bodies?
20. Are there active pastoral organisations that are dealing with the problem? Are there other organisations or people that pastoralists can turn for support in cases of conflict and crisis?

Summary, Recommendations and Conclusions

21. What is needed to alleviate human suffering, stem environmental degradation and bring about a resolution to conflicts over land?
22. Where are the gaps in knowledge that need to be filled to obtain a better understanding of the problem?

This is not meant to be a complete list. It is also accepted that questions on this list may not be relevant to every context, since each area will present variations. These differences are as important as any consistencies found. The researcher should not feel confined to the above framework. However, justification for substantive departures should be notified in advance of report preparation. It is expected that the research will be prepared mainly from the researcher's own knowledge and a review of existing materials. Where possible this is to be supplemented by fieldwork, although it is acknowledged that enquiries are limited by the time and resources made available for this work. Reference to fields of inadequate knowledge are to be declared rather than avoided.

REPORTING

It is expected that the research report will follow the following format:

- Introduction

- Policy/law context

- Current and past trends in pastoral development/welfare

- Main areas of alienation/conflict

- Environmental implications

- Case material

- Procedures for conflict resolution

- Summary/recommendations/conclusions

Preparation of a draft paper is expected within six (6) months of undertaking the work. Report length should not exceed 40 pages of double spaced type. Appendices of relevant documents and maps are welcomed. Case material should be used where possible.

APPENDIX I

Conclusions and recommendations should be drawn from the material presented in the report. Please address all correspondence to: Dr Charles Lane, IIED, 3 Endsleigh Street, London WC1H ODD, UK. Telephone: + 44 171-388 2117, fax: + 44 171-388 2826, email: IIEDDrylands@gn.apc.org

Appendix II

INTERNATIONAL ORGANISATIONS WORKING ON PASTORALIST ISSUES

Norwegian Church Aid
PO Box 4544
Torshov
N-0404 Oslo
Norway
Tel.   +47 22 22 22 99
Fax   +47 22 22 24 20

Arid Lands Information Network (ALIN)
Reseau d'Information des Terres Arides (RITA)
Casier Postal 3
Dakar-FANN
Senegal
Tel.   +221 251808
Fax   +221 254521

Publications: *Baobab, Development Projects in Arab Lands* series

Arid Lands and Resource Management Network
Centre for Basic Research
PO Box 9863
Kampala
Uganda
Tel.   +256 41 231 228
Fax   +256 41 235 413
Telex   61522

Publications: *Research Reports*

Commission on Nomadic Peoples
International Union of Anthropological and Ethnographical Sciences
c/o Institut für Volkerkunde
Universität zu Köln
Albertus-Magnus-Platz
D-50923 Köln
Germany
Tel.   +49 221 470 2278
Fax   +49 221 470 5117
E-mail alv04@rs1.rrz.uni-koeln.de

Publications: *Nomadic Peoples*

Drylands Programme
International Institute for Environment & Development (IIED)
3 Endsleigh Street
London WC1H ODD
United Kingdom
Tel.   +44 171 388 2117
Fax   +44 171 388 2826
E-mail drylands@iied.org

Publications: *Haramata, Issues Papers, Pastoral Land Tenure Series Papers, Programme Reports*

Environmental Policy and Society (EPOS)
Sturegatan 9, 1 tr
S-753 14 Uppsala
Sweden
Tel.   +46 18 18 33 25
Fax   +46 18 18 27 32 / 31 42 92
Telex   8195077
E-mail anders.hjort@epos.uu.se

Publications: *Dryland Working Paper Series*

Forests, Trees and People Programme
Department of Rural Development Studies
Swedish University of Agricultural Sciences
PO Box 7005
S-75007 Uppsala

Sweden
Tel.   +46 18 67 21 04
Fax   +46 18 67 34 20
E-mail   Bitte.Linder@irdc.slu.se

Publications: *Newsletter, Working Papers*

Centre International de Hautes Etudes Agronomiques Méditerranéennes (CIHEAM)
3191 Route de Mende
BP 5056
34033 Montpellier Cedex 1
France
Tel.   +33 67 04 60 00
Fax   +33 67 54 25 27

Publications: *Parcours demain*

Institute for Development Studies
University of Sussex
Brighton
BN1 9RE
United Kingdom
Tel.   +44 1273 606261
Fax   +44 1273 621202
E-mail   ids@sussex.ac.uk

Publications: *Discussion Papers, Bulletins, Working Papers, Annotated Bibliographies*

League for Pastoral Peoples
Pragelatostrasse 20
6432 Ober-Ramstadt
Germany
Tel./fax   +49 6154 53642

Publications: *Newsletter*

Lutheran World Relief
390 Park Avenue South
New York

New York 10016
USA
Tel.   +1 212 532 6350
Fax   +1 212 213 6081

Pastoral Development Network
ODI
Portland House
Stag Place
London SW1E 5DP
United Kingdom
Tel.   +44 171 393 1600
Fax   +44 171 393 1699
E-mail   odi@odi.org.uk

Publications: *Newsletter, PDN Network Papers*

Pastoral and Environmental Network for the Horn of Africa (PENHA)
PO Box 4014
1 Laney House
Portpool Lane
London EC1N 7UL
United Kingdom
Tel.   +44 171 242 0202
Fax   +44 171 404 6778
E-mail   penha@ukonline.co.uk

Publications: *Newsletter, Reports, Events*

Pastoral Information Network Programme (PINEP)
University of Nairobi
Department of Range Management
PO Box 29053
Kabete
Kenya
Tel.   +254 2 63 01 81
Fax   +254 2 63 12 26
Telex   22095 VARSITY KE
E-mail   pinep@attmail.com

Publications: *Working Paper Series*

Pastoral Steering Committee
c/o Oxfam
PO Box 40680
Nairobi
Kenya
Tel.   +254 2 442122
Fax   +254 2 442123

Publications: *The Pastoralist*

Scandinavian Institute of African Studies
PO Box 1703
S-751 47 Uppsala
Sweden
Tel.   +46 18 15 54 80
Fax   +46 18 69 56 29

School of Development Studies
University of East Anglia
Norwich NR4 7TJ
United Kingdom
Tel.   +44 1603 456 161
Fax   +44 1603 451 999
E-mail   odg.gen@uea.ac.uk

Vétérinaires Sans Frontières
12 rue Mulet
69001 Lyon
France
Tel.   +33 4 78 69 79 59

INTERNATIONAL DEVELOPMENT NGOS WORKING WITH PASTORALISTS IN AFRICA
(MOST HAVE COUNTRY OR REGIONAL OFFICES)

ACORD
Dean Bradley House
52 Horseferry Road

London SW1P 2AF
United Kingdom
Tel.   +44 171 227 8600
Fax   +44 171 799 1868
E-mail acord@gn.apc.org

Publications: *Of Cattle and Camels*

African Initiatives
41 Ashgrove Road
Bristol BS7 9LF
United Kingdom
Tel.   +44 117 952 0988
E-mail mikesansom@gn.apc.org

Publications: *Newsletter, Information Sheets*

Action Aid
Hamlyn House
MacDonald Road
London N19 5PG
United Kingdom
Tel.   +44 171 281 4101
Fax   +44 171 281 5146
E-mail mail@actionaid.org.uk

CAFOD
2 Romero Close
Stockwell Road
London SW9 9TY
United Kingdom
Tel.   +44 171 733 7900
Fax   +44 171 274 9630
Telex   893347 CAFOD G
E-mail reception@cafod.org.uk

Care UK
Dudley House
36–38 Southampton Street
London WC2E 7HE

United Kingdom
Tel.   +44 171 379 5247
Fax   +44 171 379 0543
E-mail   harvey@uk.care.org

CUSO
2255 Carling Avenue, Suite 400
Ottawa
Ontario K2B 1A6
Canada
Tel.   +1 613 829 7445
Fax   +1 613 829 7996
E-mail   cusoa@web.apc.org

FARM Africa
9–10 Southampton Place
London SW9 9TY
United Kingdom
Tel.   +44 171 430 0440
Fax   +44 171 430 0460
E-mail   farmafricauk@gn.apc.org

HIVOS
Raamweg 16
2596
HL Den Haag
The Netherlands
Tel.   + 31 70 3636907
Fax   + 31 70 3617447
Telex   716000

Intermediate Technology
Myson House
Railway Terrace
Rugby CV21 3HT
United Kingdom
Tel.   +44 1788 560631
Fax.   +44 1788 540270
Telex   317210 BUREAU G
E-mail: itdg@gn.apc.org

Mellemfolkeligt Samvirke (MS)
Borgergade 14
Copenhagen K
Denmark
Tel.   +45 3332 6244
Fax   +45 3315 6243

Natural Peoples World
Dronningsgade 14
DK-1420 Copenhagen K
Denmark
Tel.   +45 3154 3342
Fax   +45 3154 3392

Novib
PO Box 30919
2500 GX The Hague
The Netherlands
Tel.   +31 70 342 16 21
Fax   +31 70 361 44 61
E-mail   admin@novib.antenna.nl

Oxfam
274 Banbury Road
Oxford OX2 7DZ
United Kingdom
Tel.   +44 1865 311 311
Fax   +44 1865 312 600
E-mail   oxfam@oxfam.org.uk

SOS Sahel
1 Tolpuddle Street
London N1 OXT
United Kingdom
Tel.   +44 171 837 9129
Fax   +44 171 837 0856
E-mail   sossaheluk@gn.apc.org

INTERNATIONAL INDIGENOUS HUMAN RIGHTS ORGANISATIONS

African Rights
11 Marshalsea Road
London SE1 1EP
United Kingdom
Tel.   +44 171 717 1224
Fax   +44 171 717 1240
E-mail   afrights@gn.apc.org

Committee for Pastoralist Issues
Box 15, 8585 Glesborg
Denmark
Tel.   +45 3157 3193
Fax   +45 3296 9564

FIAN
PO Box 102243
D-69012 Heidelberg
Germany
Tel.   +49 6221 830620
Fax   +49 6221 830545

Human Rights Watch/Africa
485 Fifth Avenue
New York
NY 10017-6104
USA
Tel.   +1 212 972-8400
Fax   +1 212 972-0905
E-mail   hrwnyc@hrw.org

The International Working Group for Indigenous Affairs (IWGIA)
Fiolstaede 10
DK-1171 Copenhagen K
Denmark
Tel.   +45 33 12 47 24
Fax   +45 33147749
E-mail   iwgia@login.dkuug.dk

Minority Rights Group
379 Brixton Road
London SW9 7DE
United Kingdom
Tel.   +44 171 978 9498
Fax   +44 171 738 6265
E-mail   minority.rights@mrg.sprint.com

Society for Threatened Peoples
PO Box 2024
D-37010 Gottingen
Germany
Tel.   +49 551 49906-0
Fax   +49 551 58028

Survival International
11–15 Emerald Street
London WC1N 3QL
United Kingdom
Tel.   +44 171 242 1441
Fax   +44 171 242 1771
E-mail   survival@gn.apc.org

TROCAIRE
169 Bootenstown Avenue
Blackrock
County Dublin
Ireland
Tel.   +353 1 288 5385
Fax   +353 1 288 3577

UN Agencies
c/o Centre for Human Rights
UN Office at Geneva
8–14 avenue de la Paix
1211 Geneva 10
Switzerland
Tel.   +41 22 907 1234
Fax   +41 22 917 0123

INDIGENOUS PASTORALIST NGOS IN AFRICA

East Africa

Bulgalda (Barabaig)
PO Box 146
Katesh
Arusha
Tanzania

Dupoto e-Maa (Maasai)
Olkajiado Development Programme
PO Box 213
Kajiado
Kenya

Group Ranch Education Programme (Maasai)
PO Box 4
Loitokitok
Kenya

Inyuat e-Maa (Maa speakers)
PO Box 2720
Arusha
Tanzania

Ilaramatak Lorkonerei (Maasai)
PO Box 12785 Meru Branch
(Orkesumet/Terrat)
Arusha
Tanzania

Imusot e Purka
PO Box 17
Handemi
Tanzania

Inyuat e-Moipo (Maasai)
PO Box 519

Arusha
Tanzania

KINNAPA (Maasai)
PO Box 83
Kibaya
Arusha
Tanzania

KIPOC (Maasai)
PO Box 94
Loliondo
Arusha
Tanzania

Loita Naimina Enkiyio Conservation Trust (Maasai)
PO Box 392
Narok
Tanzania

Maa Development Association (Maasai)
PO Box 230
Narok
Kenya

Ngorongoro Crater Pastoralist Survival Trust (Maasai)
PO Box 29
Ngorongoro
Arusha
Tanzania

Ngorongoro Pastoralist Development Organisation (Maasai)
PO Box 2611
Arusha
Tanzania

Organisation for Survival of Illaikipiak Indigenous Maasai Group Initiatives (OSILIGI)
PO Box 68
Dol Dol via Nanyuki

Kenya

Oseremi (Maasai)
PO Box 70
Lonondo
Tanzania

PINGOs (coalition of indigenous pastoralist NGOs)
PO Box 12785
Arusha
Tanzania

West Africa

APESS
BP 291
Dori
Burkina Faso

AREN
BP 12669
Niamey
Niger
Tel.   227 740671

Association Pastorale de Karwassa
s/c NEF
Cercle de Douentza
Région de Mopti
Mali

Groupement d'Artisans Ruraux d'Intadayni
BP 12713
Niamey
Niger

Groupement National des Associations Pastorales en Mauritanie
BP 604
Nouakchott
Mauritania

SECADEV
BP 1166
N'Djamena
Chad

TASSAGHT
BP 32
Gao
Mali

Union de Solidarité et l'Entre-aide
BP 5070
Dakar-Fann
Senegal
Tel.   + 221 24 67 96
Fax   + 221 24 19 89

RESEARCH ORGANISATIONS IN AFRICA WORKING ON PASTORALIST ISSUES

East Africa

African Centre for Technology Studies (ACTS)
PO Box 45917
Nairobi
Kenya
Tel.   +254 2 565173/569986
Fax   +254 2 569989
E-mail   acts@elci.gn.apc.org

Publications: *Drylands Research Series*

Centre for Basic Research (CBR)
PO Box 2875
Kampala
Uganda
Tel.   +256 41 242987/231228
Fax   +256 41 235413/245580
E-mail cbr@mukla.gn.apc.org

Publications: *Working Papers*

Land Rights Research and Resources Institute (LARRRI)
PO Box 75885
Dar es Salaam
Tanzania
Tel.   +255 51 152448
Fax   +255 51 152448
E-mail ishivji@ud.co.tz

West Africa

Associates in Research and Education for Development (ARED)
BP 10737
Dakar-Liberté
Senegal
Tel.   +221 257119
Fax   +221 247097
E-mail ared@enda.sn

Publications: *Five book series in national languages (especially Pulaar) covering the following topics:*

Module 1: *Basic Literacy and Math Skills*
Module 2: *Planning and Leadership Skills*
Module 3: *Civic Society (Legal Education, Current Events & Issues)*
Module 4: *Traditional Knowledge Systems*
Module 5: *Scientific and Technical Information*

Centre de Suivi Ecologique
BP 15532
Dakar-Fann

Senegal
Tel.   +221 25 80 66
Fax   +221 25 81 68
E-mail   Aicha@cse.cse.sn

CILSS
BP 7049
Ouagadougou 03
Burkina Faso
Tel.   +226 30 67 58/30 67 59
Fax   +226 31 19 82

Institut Sénégalais de Récherches Agricoles
BP 3120
Parc de Hann
Dakar
Senegal
Tel.   +221 32 24 31
Fax   +221 32 24 27

PRASET
01 BP 1485
Ouagadougou 01
Burkina Faso
Tel.   +226 30 88 60
Fax   +226 31 25 43

REFERENCES

Introduction

Abel, N and Blaikie, P (1990) 'Land Degradation, Stocking Rates and Conservation Policies in the Communal Rangelands of Botswana and Zimbabwe' Pastoral Development Network Paper 29a, Overseas Development Institute, London

Behnke, R (1985) 'Openrange Management and Property Rights in Pastoral Africa: A case of spontaneous range enclosure in south Darfur, Sudan' Pastoral Development Network Paper 20f, Overseas Development Institute, London

Behnke, R (1991) 'Economic Models of Pastoral Land Tenure' in the proceedings of the International Rangeland Development Symposium, Department of Range Science, College of Natural Resources, Utah State University, Logan

Behnke, R (1992) 'New Directions in Africa Range Management Policy' Pastoral Development Network Paper No 32c, Overseas Development Institute, London

Behnke, R (1994) 'Natural Resource Management in Pastoral Africa' Commonwealth Secretariat, Overseas Development Institute and International Institute for Environment and Development (IIED), London

Behnke, R and Scoones, I (1992) 'Rethinking Rangeland Ecology' Dry land Issues Paper 33, IIED, London

Birgegard, L-E (1993) 'Natural Resource Tenure: A review of issues and experiences with emphasis on Sub-Saharan Africa' Rural Development Studies No 3, Swedish University of Agricultural Sciences/International Rural Development Centre, Uppsala

Bonfiglioli, A (1992) 'Pastoralists at a Crossroads: Survival and Development Issues in African Pastoralism', report for UNICEF/UNSO Project for Nomadic Pastoralists in Africa

References

Bradbury, M, Fisher, S and Lane, C (1995) 'Working with Pastoralist NGOs and Land Conflicts in Tanzania: A report on a workshop in Terrat, Tanzania 11—15 December, 1994' Pastoral Land Tenure Series No 7, IIED, London

Bromley, D and Cernea, M (1989) 'The Management of Common Property Natural Resources: Some conceptual and operational fallacies' World Bank Discussion Paper 57, The World Bank, Washington DC

Bruce, J (1986) 'Land Tenure Issues in Project Design and Strategies for Agricultural Development in Sub-Saharan Africa' Land Tenure Centre Paper 128, Land Tenure Centre, Wisconsin

Feder, G, Onchan, T, Chalamwong, Y and Hongladarom, C (1988) *Land Policies and Farm Productivity in Thailand* World Bank Research Publication, Johns Hopkins University Press, Baltimore and London

Galaty, J (1980) 'The Maasai Group-Ranch: Politics and Development in an African Pastoral Society' in Salzman, P (ed) *When Nomads Settle* Bergin, Praeger

Galaty, J, Hjort af Ornas, A, Lane, C and Ndagala, D (1994) 'The Pastoral Land Crisis: Tenure and Dispossession in Eastern Africa' in *Nomadic Peoples* no 34/35

Hardin, G (1968) 'The Tragedy of the Commons' in *Science* vol 162, no 3859, pp1243—1248

Hardin, G (1988) 'Commons Failing' in *New Scientist* 22 October

Horowitz, M and Jowkar, F (1992) 'Pastoral Women and Change in Africa, the Middle East, and Central Asia', a report from the 'Gender Relations of Pastoral and Agropastoral Production' project, Institute for Development Anthropology, Binghampton

Hulme, M (1992) 'Rainfall Changes in Africa: 1931—60 to 1961—90' in *International Journal of Climatology* vol 12, pp685-699

Johansson, L (1991) 'Land Use Planning and the Village Titling Programme Land Policy: The case of Dirma village in Hanang district', paper presented at the Land Tenure Workshop, Arusha

Kjaerby, F (1979) 'The Development of Agro-pastoralism among the Barabaig in Hanang District' BRALUP Research Paper No 56, University of Dar es Salaam, Dar es Salaam

Lane, C (1990) 'Barabaig Natural Resource Management: Sustainable Land Use Under Threat of Destruction' UNRISD Discussion Paper 12, United Nations Research Institute for Social Development, Geneva

Lane, C (1996) *Pastures Lost: Barabaig Economy, Resource Tenure and the Alienation of their Land in Tanzania* Initiative Publishers, Nairobi

Lane, C and Moorehead, R (1994) 'New Directions in Rangeland and Resource Tenure and Policy' in *Living with Uncertainty: New Directions in Pastoral Development in Africa* by Scoones, I (ed) Intermediate Technology Publications, London

Lane, C and Scoones, I (1993) 'Barabaig Natural Resource Management' in Young, M D and Solbrig, O T (eds) *The World's Savannas, Man and Biosphere Series, Vol 12* UNESCO and Parthenon

Lane, C and Swift, J (1989) 'East African Pastoralism: Common Land, Common Problems', *Drylands Issues Paper No 8* Drylands Programme, IIED, London

Lund, C (1993) 'Waiting for the Rural Code: Perspectives on a Land Tenure Reform in Niger', *Drylands Issues Paper No 44* IIED, London

Oxby, C (1981) *Group Ranches in Africa* FAO, Rome

Place, F and Hazell, P (1993) 'Productive Effects and Indigenous Land Tenure Systems in Sub-Saharan Africa' in *American Journal of Agricultural Economics* February, pp10—19

Runge, C (1981) 'Common Property Externalities: Isolation, assurance and resource depletion in a traditional grazing context' in *American Journal of Agricultural Economics* No 63, pp595—606

Runge, C (1984) 'Institutions and the Free Rider: The assurance problem in collective action' in *Journal of Politics* vol 46, pp154—181

Runge, C (1986) 'Common Property and Collective Action in Economic Development', National Research Council, Proceedings of the Conference on Common Property Resource Management, National Academy Press, Washington DC

Sandford, S (1983) *Management of Pastoral Development in the Third World* Overseas Development Institute, London

Scoones, I (1991) 'Wetlands in Drylands: The Agroecology of Savanna Systems in Africa', *Drylands Programme Report* IIED, London

Thebaud, B (1990) 'Politiques Hydrauliques Pastorales et Gestion de L'espace au Sahel', in *Societes pastorales et developpement: Cahiers des sciences humaines* vol 26 (1-2): 13—21

Thebaud, B (1993) Contribution to the Research Workshop on New Directions in African Range Management and Policy, Woburn

Toulmin, C (1994) "Gestion de Terrior": Concepts and Development' UNSO, New York

White, R (1992) 'Livestock Development and Pastoral Production on Communal Rangeland in Botswana', Food Production and Rural Development Division, Commonwealth Secretariat, England

World Bank (1989) *Sub-saharan Africa: From Crisis to Sustainable Growth* The World Bank, Washington DC

Zeidane, M ould (1993) 'Pastoral Associations: Recent Evolution and Future Perspectives', contribution to Research Workshop on New Directions in African Range Management and Policy, Woburn, England

Kenya

Atwood, D (1990) 'Land Registration in Africa: The impact on agricultural production' in *World Development* vol 18, no 5, pp659—71

Berger, J (1989) 'Wildlife Extension, A Participatory Approach to Conservation: A Case Study Among the Maasai People of Kenya' University of California, Berkeley, PhD thesis

Government of Kenya (1986a) *Narok District Socio-Cultural Profile* Institute of African Studies, University of Nairobi

Government of Kenya (1986b) 'Economic Management for Renewed Growth' Sessional Paper No 1

Government of Kenya (1986c) *Statistical Abstracts for 1980—1988* Nairobi, Government Printer

Grandin, B, De Leeuw, P and Lembuya, P (1989) 'Drought, Resource Distribution and Mobility on Two Maasai Group Ranches, Southeastern Kajiado District' in Downing, T et al *Coping with Drought in Kenya* Lynne Rienner Publishers, Boulder, Colorado

Helland, J (1987) *Turkana Briefing Notes—a Background Study of the Turkana Rural Development Programme* The Christian Michelsens Institute, Bergen

Hjort af Ornäs, A (1990) 'Environment and the Security of Dryland Herders' in Hjort af Ornäs, A and Mohamed Salih, M (eds) *Ecology and Politics: Environmental Stress and Security in Africa* Uppsala, Scandinavian Institute of African Studies

Hogg, R (1986) 'The New Pastoralism: Poverty and Dependency in Northern Kenya' in *Africa* 56(3), pp319—333

Juma, C (1989) 'Environment and Economic Policy in Kenya' in Kiriro, A and Juma, C (eds) *Gaining Ground: Institutional Innovations in Land-use Management in Kenya* ACTS Press, Nairobi

Kipuri, N (1989) 'Maasai Women in Transition: Class and Gender in the Transformation of a Pastoral Society' Department of Social Anthropology, Temple University, PhD thesis

Kitching, G (1980) *Class and Economic Change in Kenya: The Making of an African Petite-Bourgeoisie* Yale University Press, New Haven

Kituyi, M (1990) *Becoming Kenyans: Socio-Economic Transformation of the Pastoral Maasai* ACTS Press, Nairobi

Kituyi, M (1985) *The State and The Pastoralists: The Marginalization of the Kenyan Maasai* The Christian Michelsens Institute, DERAP Working Paper, Bergen

Lamprey, H and Yussuf, H (1981) 'Pastoralism and Desert Encroachment in Northern Kenya' in *Ambio*, vol 10, No 2—3

Lane, C (1990) 'Barabaig Natural Resource Management: Sustainable Land Use Under Threat of Destruction' UNRISD Discussion Paper 12

Lane, C (1996) *Pastures Lost: Barabaig Economy, Resource Tenure and the Alienation of their Land in Tanzania* Initiatives Publishers, Nairobi

Leys, C (1977) *Underdevelopment in Kenya: The Political Economy of Neo-colonialism 1964—1971* Heinemann, London

Livingstone, I (1977) 'Economic Irrationality among Pastoral Peoples: Myth or Reality?' in *Development and Change*, vol 8, pp209—30

Loiske, V M (1990) 'Political Adaption: The Case of the Wabarabaig in Hanang District, Tanzania' in Bovin and Manger (eds) *Adaptive Strategies in African Arid Lands* Scandinavian Institute for African Studies, Uppsala, pp77—90

Okoth-Ogendo, H (1976) 'African Land Tenure Reform' in Heyer, Maitha and Senga (eds) *Agricultural Development in Kenya: An Economic Assessment* Oxford University Press, Nairobi

Okoth-Ogendo, H W O (1991) *Tenants of the Crown: Evolution of Agrarian Law and Institutions in Kenya* ACTS Press, Nairobi

World Commission on Environment and Development (1987) *Our Common Future* Oxford University Press, Oxford

Raintree, J (1987) 'Land Trees and Tenure' proceedings of an international workshop on tenure issues in agroforestry, ICRAF, Nairobi

Swynerton, R (1955) 'A Plan to Intensify African Agriculture in Kenya' Government Printer, Nairobi

White, J and Meadows, S (1981) 'Evaluation of the Contribution of Group and Individual Ranches in Kajiado District, Kenya' Ministry of Livestock Development, Nairobi

Mali

Barth, H (1890) 'Travels and Discoveries in North and Central Africa' in *The Minerva Library of Famous Books* Bettany, G T (ed) London, Ward, Lock and Co

Behnke, R H and Scoones, I (1991) *Rethinking Range Ecology: Implications for Rangeland Management in Africa* Overseas Development Institute and IIED, London

Blaikie, P M and Brookfield, H (1987) *Land Degradation and Society* Methuen, London

CIPEA (Centre International pour l'Elevage en Afrique) (1983) *Recherche d'une Solution aux Problèmes de l'Elevage dans le Delta Intérieur du Niger au Mali: Volume 5, Rapport de Synthèse*, CIPEA and ODEM, Bamako, Mopti and Addis Ababa, mimeo

Gallais, J (1967) 'Le Delta Intérieur du Niger: Etude de Géographie Régionale' in *Mémoires de l'Institut Fondamental d'Afrique Noir* no 79, IFAN, Dakar, mimeo

Hardin, G (1968) 'The Tragedy of the Commons' in *Science*, no 162, pp1243—1248, reprinted in Hardin, G and Baden, J (eds) *Managing the Commons* W H Freeman and Co, San Francisco

IUCN (1987) 'Conservation de l'Environnement dans le delta Intérieur du Fleuve Niger: Document de Synthèse', Projet de Conservation dans le delta Intérieur du Niger, Rapport Technique No 3, MRNE, Direction Nationale des Eaux et Forêts, Mopti and IUCN, Gland, mimeo.

IUCN (1989) 'Rapport sur l'Economie de la Région de Mopti (Mali) 1970–1985', Programme Sahel, IUCN, Gland, mimeo

Moorehead, R M (1991) 'Structural Chaos: Community and State Management of Common Property in Mali', unpublished PhD Thesis, IDS, University of Sussex

Oakerson, R J (1986) 'A Model for the Analysis of Common Property Problems' in *National Research Council Proceedings of the Conference on Common Property Resource Management*, pp13—29, National Academy Press, Washington, DC

République du Mali (1987a) *Code Domanial et Foncier*, Bamako

République du Mali (1987b) *Le Plan National de Lutte contre la Désertification et l'Avancée du Désert*, Ministère des Ressources Naturelles et de l'Elevage, Bamako, mimeo

République du Mali (nd) *Textes Forestières*, Ministère des Ressources Naturelles et de l'Elevage, Direction Nationale des Eaux et Forêts, Bamako, mimeo

République du Mali et Confédération Suisse (1987) *Rapport de la Mission Conjointe d'Etude de la Police Forestière*, Ministère des Ressources Naturelles et de l'Elevage et Direction de la Coopération au Développement et de l'Aide Humanitaire, Bamako, mimeo

Shanmugaratnam, N, Vedeld, T, Mossige, A and Bovin, M (1992) 'Resource Management and Pastoral Institution Building in the West African Sahel' World Bank Discussion Papers, Africa Technical Department Series, No 175, The World Bank, Washington, DC

Turner, M D (1992) 'Life on the Margin: Fulbé Herding Practices and the Relationship between Economy and Ecology in the Inland Niger Delta of Mali', unpublished PhD dissertation, University of California at Berkeley

Mauritania

Bonte, P (1987) 'L'herbe ou le sol L'evolution du systeme foncier pastoral en Mauritanie du Sud Ouest' in Gast, M (ed) *Heriter en pays Musulman* CNRS, Aix-en-Provence

Bonte, P (1983) 'Esquisse d'histoire fonciere de l'Emirat de l'Adrar' in Baduel, R (ed) *Etats, Territoire et Terroirs au Maghreb*, Ed. du CRNS, Extrait de l'Annuaire de l'Afrique du Nord, Paris

Bonte, P (1990) Project Elevage II, Rapport de Synthèse, Programme de Généralisation des Associations Pastorale, R.I. Mauritanie, Nouakchott

Boutillier, J L and Schmitz, J (1986) *Gestion traditionnelle des terres et transition vers l'irrique, le cas de la vallee du Sénégal* Paris-PUF

Gaye, M and Zeidane, M O (1991) 'La commercialisation du Betail, de la Viande et des produits d'origine animale', CILSS, Nouakchott

SEDES (1981) 'Projet de Financement pour un projet de développement de l'élevage dans le Sud-Est Mauritanien', Tome II: *Dossier de Presentation des Opérations Préconisées,* Paris

The World Bank (1986) 'Staff Appraisal Report, Islamic Republic of Mauritania', Second Livestock Project, No 5720—MAU, Western African Projects Department, Agriculture C, Washington, DC

Senegal

Ba, C (1982) 'Les peul du Sénégal: étude géographique', Université de Paris VII, Thèse de doctorat d'Etat et lettres et Sciences Humaines

Baker, R (1979) 'Perception de l'état pastoral' in *Environnement Africain* Série Etudes et Recherches No 46—79, ENDA, Dakar

Barral, H (1982) *Le Ferlo des forages—Gestion actuelle et ancienne de l'espace pastoral,* Orstom, Dakar

Barral, H, Benefice, E, Boudet, G et al (1983) 'Systèmes de production d'élevage au Sénégal dans la région du Ferlo—Synthèse de fin d'étude d'une équipe pluridisciplinaire' Acc-Griza (Lat), Maisons-Alfort, IEMVT

Bille, J C (1991) 'Un pastoraliste au Kenya in Pastum' in *Bulletin de l'association Française de Pastoralisme* 22, pp16—17

Bugnicourt, J (1977) 'La zone du Sahel: un environnement difficile et précaire', Textes choisis ENDA Tiers Monde

Dieme, I (1986) Présentation des activités du PDESO—Acte du séminaire InterEtats sur les associations Pastorales tenu à Tambacounda du 22 février au 1er mars 1986, PDESO, Tambacounda, pp11—24

Guèye, I S (1985) 'De la participation des usagers aux charges d'exploitation des forages gérés par la SODESP', Note technique No 20, SODESP, Dakar

Grosmaire, P (1957) 'Eléments de politique sylvo-pastorale', Inspection forestière du fleuve, Saint Louis; Service des Eaux Forêts, 18 fascicules

Niane, I C (1987) 'Du fonctionnement des conseils ruraux et des centres d'expansion rurale polyvalents de la Vallée du fleuve Sénégal', Projet SEN/86.001—Assistance à la cellule Après-Barrages

Niane, I C, 'Fonctionnement des conseils ruraux et des centres d'expansion rurale polyvalents du Delta: les cas de Ross-Bethio et Gae', PNUD/Cellule Après Barrage, Dakar

Niane, I C (1990) 'Note sur la situation des affectations foncières (1988–1990) et les litiges foncières intra et inter-communautaires' Département de Matam, Podor et Bakel

Niane, M (1984) 'Le projet environnemental devant la dynamique agro-pastorale dans le bassin du Fleuve Sénégal: l'exemple du Galogjina', ISE/Université de Dakar; DEA en Sciences de l'Environnement

Niang, M (1982) 'Réflexions sur la réforme foncière sénégalaise de 1964' in *Enjeux fonciers en Afrique Noire*, Orstom, Kharthala, pp220—227

Penning de Wries, F W T and Djitèye, M A (1982) 'La productivité des pâturages sahéliens: une étude des sols, des végétations et de l'exploitation de cette ressource naturelle', Centre for Agricultural Publishing and Documentation, Wageningen

Pouillon, D G (1984) 'Evaluation de l'Elevage bovin dans la zone sahélienne du Sénégal' in *Etude sociologique* pp69—112

Santoir, C (1982) 'Contribution à l'étude de l'exploitation du cheptel, Région du Ferlo, Sénégal', Orstom, Dakar

Santoir, C (1983) 'Raison pastorale et politique de développement: les Peul sénégalais face aux aménagements', Travaux et documents No 166, Orstom, Paris

Touré, I A (1983) 'Gestion pastorale et développement intégré'. Situations, perspectives au Sénégal, Communication au colloque sur les méthodes d'inventaire et de surveillance continue des écosystèmes pastoraux sahéliens, PNUD/FAO-ISRA, Application au développement tenu a Dakar, 16–18 Novembre

Touré, O (1991) 'Projet Sénégalo-Allemand d'Exploitation Agro-Sylvopastorale des Sols dans le nord du Sénégal—Mission de consultation' in *Rapport Sociologique*

Wincke, P P (1990) *Evaluation des potentialités et contraintes d'une zone écologique: la zone sylvo-pastorale*, Cellule Après Barrages, Dakar

LEGISLATION AND REGULATIONS CITED

1964: Law No 64—46 of 17 June concerning state-administered property. Enforcement order No 64—573 of 30 July setting the conditions for its enforcement
1965: Decree No 65—078 of 10 February 1965 establishing the forestry code (regulatory part)
1966: Decree No 66—45 relating to the declassification of the Boulal and Deali forests
1972: Law No 72—25 of 19 April on administrative, territorial, and local reform

1974: Law No 74—46 of 18 July establishing the forestry code (legislative part)
1980: Law No 80—14 of 3 June abrogating and replacing certain articles of Law No 72—25 of 19 April 1972 concerning rural communities Decree No 80—268 of 10 March 1980 concerning the organisation of rangeland and assessing the conditions for the use of pasture
1986: Decree No 86—320 of 11 March on the regulation of animal husbandry, the introduction and use of camels in Senegal

Sudan

Bashari, O A (1985) 'Land Tenure in Rainfed Area—Sudan', World Bank, vol 4, annex VI, Livestock, Washington, DC

Lebon, J H G (1965) 'Land Use in the Sudan, the World Land Use Survey', monograph No 4, Geographical Publications

Tanzania

Arhem, K (1985a) *Pastoral Man in the Garden of Eden* University of Uppsala, Sweden

Arhem, K (1985b) *The Maasai and the State* IGWIA Document No 52, Copenhagen, Denmark

Beidelman, T O (1960) 'The Baraguyu' in *Tanzania Notes and Records* no 55, pp245—78

Bilali, A (1989) 'Management of Pastures and Grazing Lands in Tanzania' in *Splash*, vol 5, no 2, SADCC

Bovin, M and Manger, L (1990) (eds) *Adaptive Strategies in African Arid Lands* Scandinavian Institute of African Studies, Uppsala

Director of Veterinary Services (1948) 'Annual Report, Department of Veterinary Sciences and Animal Husbandry', Government Printer, Dar es Salaam

Galaty, J and Aronson, D (1981) *The Future of Pastoral Peoples* IDRC, Ottawa

Gulbrandsen, O, 'On changing implications of property rights, socio-economic inequality and scarcity in Bangwaketse Society, Botswana', paper presented at the workshop on rights in property and problems of pastoral development (Manchester, April 23—24)

Hardin, G (1968) 'The Tragedy of the Commons' in *Science* no 162, pp1243—8

Hardin, G (1986) *Filters Against Folly: How to Survive Despite Economists, Ecologists and the Merely Eloquent* Penguin, New York

Helland, J (1990) 'Finding a New Pastoralism or Sustaining Pastoralists? The dilemma facing East African pastoralists' in *Utviklingsstudier* no. 2, pp167—82

Herskovits, M (1926) 'The Cattle Complex in East Africa' in *American Anthropologist* no 28, pp230—72

Hjort af Ornas, A and Mohamed Salih, M (1989) (eds) *Ecological Degradation and Political Conflicts in Africa* Scandinavian Institute of African Studies, Uppsala

Horowitz, M (1981) Comment on Samford, 'Organizing government's role in the pastoral sector' in Galaty et al (eds) *The Future of Pastoral Peoples*, IDRC, Ottawa

James, R W and Fimbo, G M (1973) *Customary Land Law of Tanzania* East African Literature Bureau, Nairobi

Kituyi, M and Kipuri, N (1991) 'Changing pastoral land tenure and resource management in Eastern Africa: a research agenda', paper presented to workshop on land tenure and resource management among pastoralists (Nairobi, June 21—23)

Kjaerby, F (1979) *The Development of Agro-Pastoralism Among the Barabaig in Hanang District*, BRALUP Research Paper no 56, Dar es Salaam

Kjaerby, F (1980) *The Problem of Livestock Development and Villagisation Among the Barabaig in Hanang District,* BRALUP Research Paper no 40 (new series), Dar es Salaam

Kjaerby, F (1989) *Villagisation and the crisis: agricultural production in Hanang district, northern Tanzania*, Project Paper No. 89.2, Centre for Development Research, Copenhagen

Lane, C (1990a) 'Barabaig natural resource management: sustainable land use under threat of destruction', UNRISD Discussion Paper no 12, UNRISD, Geneva

Lane, C (1990b) 'Wheat at What Cost: the Tanzania Canada Wheat Program' in *Conflicts of Interest: Canadian Aid to the Third World*, Between the Lines, Toronto

Lane, C (1996) 'Pastures Lost: Barabaig Economy, Resource Tenure, and the Alienation of their Land in Tanzania', Initiatives Publishers, Nairobi

Lane, C and Pretty, J (1990) 'Displaced pastoralists and transferred wheat technology in Tanzania' in *Gatekeeper* no SA20, IIED, London

Lane, C and Scoones, I (1991) *'Barabaig natural resource management: implications for sustainable savanna use in pastoral areas of Africa'*, paper presented at IUBS/UNESCO/UNEP/CED International Symposium on Economic Driving Forces and Ecological Constraints in Savanna Land Use, Nairobi, January

Lane, C and Swift, J (1989) 'East Africa Pastoralism: common land common problem', Issue Paper no 8, IIED, London

Loiske, V B (1990) 'Political adaptation: the case of the Wabarbaig in Hanang District, Tanzania' in Bovin, M and Manger, L (eds) *Adaptive Strategies in African Arid Lands*, Scandinavian Institute of African Studies, Uppsala

Mazonde, I (1987) 'From communal water points to private wells and boreholes in Boswana's communal areas: process of individuation and socio-economic differentiation among African pastoralists', paper presented at the workshop on changing rights in property and problems of pastoral development (Manchester, April 23—24)

Mohamed Salih, M A (1990) 'Pastoralism and the state in African arid lands' in *Nomadic Peoples* nos 25—27, pp7—18.

Morris, J (1975) *'Maasai rangeland development: a progress report on the implementation of the Maasai Range Project'*, paper presented at the annual meeting of the Tanzania Society of Animal Production, Morogoro, Tanzania

Mustafa, K (1989) *Participatory Research and the 'Pastoralist Question' in Tanzania* University Press, Helsinki

Ndagala, D K (1974) *'Social and Economic Change among the Pastoral Wakwavi and its Impact on Rural Development'*, MA thesis, University of Dar es Salaam

Ndagala, D K (1978) *'Operesheni au Hatua kwa Hatua'*, working paper written for the

Rift Valley Project, Research Department, Ministry of National Culture and Youth, Tanzania

Ndagala, D K (1982) 'Operation Imparnati: the sedentarisation of the pastoral Maasai in Tanzania' in *Nomadic Peoples* no 18, pp3—8

Ndagala, D K (1986) 'The Ilparakuyo Livestock Keepers of Bagamoyo: Persistent Fighters but Ultimate Losers' in *Working Papers in African Studies* no 32 Uppsala University

Ndagala, D K (1990a) 'Pastoral territoriality and land degradation' in Palsson, G (ed) *From Water to World-Making* The Scandinavian Institute of African Studies, Uppsala

Ndagala, D K (1990b) *Territory, Pastoralists and Livestock: Resource Control Among the Kisongo Maasai*, Uppsala University

Ndagala, D K (1990c) 'Pastoralists and the state in Tanzania' in *Nomadic Peoples* nos 25—27, pp51—64

Ndagala, D K (1991a) *Pastoralism and Rural Development: The Ilparakuyo Experience* New Delhi, Reliance Publishing House

Ndagala, D K (1991b) 'The Unmaking of the Datoga: decreasing resources and increasing conlict in rural Tanzania' in *Nomadic Peoples* no 28, pp71—81

Nyerere, J K (1996) *Freedom and Unity/Uhuru na Umoga* Dar es Salaam, Oxford University Press

Oba, G (1987) *'Changing property rights among settling pastoralists: an adaptive strategy to declining resources'*, paper presented at the workshop on changing rights in property and problems of pastoral development, Department of Social Anthropology (Manchester, April 23—24)

Parkipuny, M L (1975) 'Maasai Predicament Beyond Pastoralism: a case study in socio-economic transformation of pastoralism', MA dissertation, IDS, University of Dar es Salaam

Parkipuny, M L (1977) 'The Alienation of Pastoralists in post-Arusha Declaration Tanzania' University of Dar es Salaam

Parkipuny, M L (1979) 'Some crucial aspects of the Maasai predicament' in Coulson, A (ed) *African Socialism in Practice, the Tanzania Experience* Nottingham, Spokesman, pp136—157

Parkipuny, M L (1990) 'Maasai development in historical perspective', Maasai Range Project paper, mimeo

Parkipuny, M L (1983) *'Maasai struggle for home right in the land of Ngorongoro Crater'*, paper presented at the Symposium on the Anthropology of Human Rights, XI International Congress of Anthropology and Ethnological Sciences, Quebec

Parkipuny, M L (1991) 'Pastoralism, Conservation and Development in the Greater Serengeti Region', IIED, Issue Paper no 26, London

Raikes, P (1981) *Livestock Development Policy in East Africa* Scandinavian Institute of African Studies, Uppsala

Rigby, P (1969) (ed) 'Pastoralism and prejudice: ideology and rural development in East Africa' in *Society and Social Change in Eastern Africa* Makerere University, Institute of Social Research, no 43, Kampala

Rigby, P (1985) *Persistent Pastoralists* Zed Press, London

Shivji, I and Tenga, R (1985) 'Ujamaa in court—reports on an acid test for peasant rights in Tanzania' in *Africa Events* vol I, no 12, pp18—20

Tanzania Government (1982) 'The Livestock Policy of Tanzania', Ministry of Livestock Development, Dar es Salaam

Tanzania Government (1994) *Report of the Presidential Commission of Inquiry into Land Matters: Volume 1; Land Policy and Land Tenure Structure* The Ministry of Lands, Housing and Urban Development, in cooperation with The Scandinavian Institute of Africa Studies, Uppsala

United States Agency for International Development (USAID) (1986) 'Land Tenure and Livestock Development in Sub-Saharan Africa' in *Aid Evaluation Study* no 39 USAID

Uganda

Baker, R P (1967) 'Environment Influences on Cattle Marketing in Karamoja', Occasional Paper No 5, Department of Geography, Makerere University, Makerere

Baker, R P (1975) '"Development" and Pastoral People of Karamoja, North-eastern Uganda: an Example of the Treatment of Symptoms' in *Pastoralism in Tropical Africa*

International African Institute, Oxford University Press, Oxford

Bennett, J W, Lawry, S W and Riddel, J C (1986) 'Land Tenure and Livestock Development in Sub-Saharan Africa', AID Evaluation Special Case Study No 39, Washington, DC

Beatie, J (1971) *The Nyoro State* Oxford University Press, Oxford

Cisternino, M (1979) 'Kalamoja, The Human Zoo: The History of Planning for Kalamoja with some Tentative Planning', dissertation, Centre for Development Studies, University of Wales, Swansea

Doornbos, R (1975) 'Land Tenure and Political Conflicts in Ankole, Uganda' in *The Journal of Development Studies* vol 12, no 1

Doornbos, R and Lofchie, F (1971) 'Ranching and Scheming: A Case Study of Ankole Ranching Scheme' in *The State of Nations: Constraints on Development in Independent Africa* by Michael Lofchie (ed), University of California Press, Berkeley

Dyson-Hudson, N (1958) *The Present Position of the Karimojong*, Colonial Office memo, Fr Novelli Bruno's Moroto Catholic Diocese Documentation Centre, Moroto

Dyson-Hudson, N (1962) 'Factors Inhibiting Change in an African Pastoral Society: the Karimojong of North-east Uganda' in *Transactions of the New York Academy of Sciences*, vol 11,24, pp771—801

Government of Uganda (1992) 'Production Zones and Targets 1992—1995', Ministry of Agriculture, Animal Industry and Fisheries, Government Printer, Entebbe

Government of Uganda (1993a) 'Rehabilitation and Development Plan 1993—1995/96', Volume I, Macroeconomic and Sectoral Policy, Ministry of Finance and Economic Planning, Government Printer, Entebbe

Government of Uganda (1993b) 'Land Disputes in Kasese District: A Report of the Cabinet Committee Investigating Land Disputes in Kasese District', Government Printer, Entebbe

Karugire, R S (1971) *A History of the Kingdom of Nkore in Western Uganda to 1896* Clarendon Press, London

Khiddu-Makubuya, E (1991) 'Land Law Reform and Rural Development in Uganda' in

Nsibambi, A and Katorobo, J (eds) *Proceedings of the Conference on Rural Rehabilitation and Development 1989*, Makerere University, Kampala

Kisamba-Mugerwa, W, Muwanga-Zaake, E S and Khiddu-Makubuya, E (1989) 'A Quarter Century of Individual Title: an Analysis of the Rujumbura Pilot Land Registration in Uganda and its Impact on Smallholder Agriculture', Madison Land Tenure Center Report, University of Wisconsin

Lane, C (1996) *Pastures Lost: Barabaig Economy, Resource Tenure, and the Alienation of their Land in Tanzania* Initiatives Publishers, Nairobi

Mugerwa, E B (1973) 'The Question of the Mailo Owners in Then Peasantry Society in Buganda: A Case Study of Muge and Lukonge Village', BA thesis, Department of Political Sciences, University of Dar es Salaam

Nsibambi, A R (1989) 'The Land Question and Conflict' in Rupensinghe (ed) *Conflict Resolution in Uganda* James Curry, London

Pulkol, D (1991) 'Report of the Ranch Restructuring Board', Part I, Ministry of Agriculture, Animal Industry and Fisheries, Government Printer, Entebbe

Morris, H F and Reed, J S (1996) 'Land Tenure and Political Conflict in Ankole Uganda' in *The Journal of Development Studies*, vol 12, no 1

Quam, M D (1976) 'Pastoral Economy and Cattle Marketing in Karamoja, Uganda', PhD thesis, Department of Anthropology, Indiana University

Richards, I A, Sturrock, F and Fortt (eds) (1973) *Subsistence to Commercial Farming in Present Day Buganda: An Economic and Anthropological Survey* Cambridge University Press, London

Slade, G and Weitz, K (1991) 'Uganda Environmental Issues and Options', Working Paper nos 91—93, School of Forestry and Environmental Studies, Duke University

United Nations (1988) 'Wildlife Protected Areas', Strategic Resources Planning in Uganda, vol II, United Nations Environment Programme (UNEP), Nairobi

West, W H (1964) 'Mailo System in Buganda', Government Printer, Entebbe

INDEX

absentee owners 153, 177
access
  to resources 7, 9–15, 16, 51–2, 100, 103–6
  to water 9, 13, 33–4, 88–9, 100, 103–4
ACORD 139
African Centre for Technology Studies 44
African Development Bank 79
agriculture *see* cultivation
agro-pastoralism 20, 72, 83–4, 109–10, 144–5, 178
aid 19–20, 21
  in Kenya 35
  in Mali 61
  in Mauritania 79–81
  in Senegal 99–101
  in Sudan 138–9
  in Tanzania 159–60
  in Uganda 176
alienation
  of land 17–19, 35, 134
  of livestock 63–4
Amin, Idi 183
animal husbandry 33, 34, 71, 74, 88, 102, 176
Arab Fund for Economic and Social Development 81
Arhem, K 164
'assurance problem' 8–9

Ba, C 113
Baker, R 103, 182
Behnke, R 7, 22, 23–4

Bilali, A 156
Bille, JC 118
Birgegard, L-E 7
Bonte, P 76, 87
Boran 33–4
boreholes 33–4, 100, 103–4, 118, 138, 161
Botswana 13, 161
Bromley, D 6
Bruce, J 6, 7
Bugnicourt, J 103

camels 121–2, 123, 124, 133
cattle 121–2, 123, 132
cattle raids 41, 151, 170, 184
charcoal burning 85, 86, 89, 162
CIDA 19
'Code Rurale' 12
colonialism
  in Kenya 16, 30
  in Mali 17, 53–6
  in Mauritania 72–3
  in Senegal 16–17, 113
  in Sudan 16, 134, 140
  in Tanzania 152–3
  in Uganda 16, 172–3
Condominium government 134
conflicts
  in Kenya 41, 43
  in Mali 63–6
  in Mauritania 82–4, 88–90
  resolution 43, 65–6, 88–90, 113–16, 147–8, 164–6, 183–4

in Senegal 20, 102–6, 110–16
in Sudan 124, 139–41, 147–8, 149
in Tanzania 160–2, 164–7
tribal 84, 124, 140–1, 147, 164–5, 166–7
in Uganda 178–80, 183–4
water 41, 88–9, 110
conservation
  in Kenya 16, 31, 117
  in Mali 67, 69
  and pastoralists 16, 31, 163, 175–6, 179, 183
  in Senegal 98–9, 117
  in Tanzania 162, 163
  in Uganda 16, 175–6, 176, 179, 184
cooperatives 75–6, 179
  herder 115–16
  women 81–2
Crown land 16, 30, 134, 152, 172
cultivation
  commercial 6, 37, 40, 113, 159, 163
  and the environment 40, 85–6, 108, 109–10, 143, 144, 162–3
  and pastoralism 64, 88–9, 109–12, 136–7, 155–6, 160–3, 175–6

dams 34, 35, 73, 87, 110, 180
DDC 159–60
deforestation 162
degradation
  of the environment 40, 85–6, 106–10, 142–5, 162–3, 180–1
  of land 34, 138
desertification 59, 85, 138
development policy
  in Kenya 33–7
  livestock 79–81, 99–102, 157–8, 159
  in Mali 53–63
  in Mauritania 74–81, 91–2
  in Senegal 96, 98–102
  in Sudan 137–9

in Tanzania 158–60
in Uganda 175–7
Diama dam 73, 87, 110
Dina system 17, 50, 51–3, 55
donors
  failure of 20
  in Kenya 35
  in Mali 61
  in Mauritania 79–81
  and pastoralists 19–20, 25
  projects 35, 79–81, 147–8, 159–60, 176
  in Senegal 99–101
  in Sudan 138–9
  in Tanzania 159–60
  in Uganda 176
drought 85, 107–8, 124, 139, 143–4, 180

economy and pastoralism 74, 136, 146, 150–1, 164, 184
enclosures *see* privatisation
environment
  and cultivation 40, 85–6, 108, 109–10, 143, 144, 162–3
  degradation 40, 85–6, 106–10, 142–5, 162–3, 180–1
  and livestock 66–7, 107–8, 109
  and sedentarisation 11, 86, 143, 180
  and water sources 20, 66–7, 85, 143, 177, 181
erosion 106, 107, 108, 163, 179, 180, 181
European Development Fund 80
evictions 135, 173, 179, 183

FAO 19, 35
farming *see* cultivation
Fimbo, G M 151
fiscal policy 57, 58–63, 67, 172
fishing 47
food
  for animals 34, 67, 85, 105, 106

production 35, 37, 60, 74, 77, 88
forestry policy
  in Kenya 34
  in Mali 60, 62, 69
  in Mauritania 76
  in Senegal 104–5, 111–12
French Aid Cooperation 80
fuel 85, 86, 89, 162, 182
Fulani *see* Fulbé
Fulbé 49–51, 63–5

game reserves
  in Kenya 16, 31
  in Senegal 99
  in Tanzania 163
  in Uganda 16, 176, 179, 183, 184
German GTZ Agency 80–1
'Gestion de Terrior' 11–12
Ghana 14
GIEs 98, 100, 116
goats 121–2, 123, 133
government
  agencies 61–3, 81, 100–2, 174, 175, 176
  bribery 64
  fiscal policy 67, 171
  intervention 34–5, 173
  investment 61–2, 63, 99
  and land policy 29–32, 58–63, 67–70, 73–9, 94–6, 98–101, 183–4
  and land tenure 16–17, 29–31, 50–8, 72–3, 151–8, 165–6, 170–5
  revenue 60, 61–2, 63, 172
  in Uganda 183
grasses 67, 106
grazing rights 31, 33, 76, 104–5, 179, 180–1, 183
Grosmaire, P 105
groundnut growing 109, 111

Helland, J 165

herders
  changing lifestyle 3–4, 63–4, 88–9
  cooperatives 115–16
  resource use 8–9, 21, 57, 66–7, 75
  vulnerability 37, 79, 91
Horowitz, M 168
hunting 31, 176

IIED 2
immigration 40–1, 51, 178–9, 180 *see also* infiltration
independence
  in Kenya 31, 32
  in Mali 56–8
  in Mauritania 73
  in Sudan 134, 140–1
  in Tanzania 153–5
  in Uganda 173–5
industry 21, 37, 40, 47, 73, 177
infiltration 31, 35, 64–5, 98, 166 *see also* immigration
inheritance 39
International Development Agency 79
investment 61–2, 63, 74, 99, 161
irrigation 60, 70, 74

James, RW 151

Kenya
  aid 35
  animal husbandry 33, 34
  case studies 40–1, 118
  colonialism 16, 30–1
  conflicts 41, 43
  conservation 16, 31, 117
  development policy 33–7
  and donors 35
  and the environment 20, 40
  food production 35, 37
  forestry policy 34

game reserves 16, 31
independence 31, 32
industry 37
land policy 29–31
land tenure 14, 16, 19, 20, 33–7
livestock ownership 36
migration 30–1
national parks 16, 31, 117
privatisation 36–7
rainfall 26
ranch system 13, 32, 36–7, 37–40
social issues 35–6, 38–40
tribes 29, 33–7
water supply 33–4
women 39
Khelcom forest 111–12
Kjaerby, F 163
Kuwaiti Development Fund 81

land
  Crown ownership 16, 30, 134, 152, 172
  degradation 34, 138
  freehold 172, 173, 174
  and government 29–32, 58–63, 67–70, 73–9, 94–6, 98–101, 183–4
  leasehold 141, 155, 172, 173, 174
  legislation 98–9, 101, 104–5, 113–16, 133–7, 166, 172–5
  registration 11–13, 44, 56–7, 167–8, 174, 179
  subdivision 13, 32, 36–7, 39
  titling 11–13, 44, 56–7, 167–8, 174, 179
  unoccupied 16, 30, 94, 134, 152, 172
  use 1–2, 8, 10–13
land tenure
  in changing context 1–2, 9–15, 21–5
  customary 12–13
  defined 6–9
  and government 16–17, 29–31, 50–8, 72–3, 151–8, 165–6, 170–5

insecurity 9–15, 54, 57, 173
in Kenya 16, 19, 20, 33–7
legislation 11–13, 17–19, 56–7, 73–7, 89, 116
in Mali 17, 50–8
in Mauritania 18, 20, 72–3, 78–9, 86–8
reform 5–6, 9–15, 172, 174
in Senegal 16–17, 18, 94, 110–13, 116
in Sudan 16, 122–3, 124, 133–7, 141, 146, 148–9
in Tanzania 19, 151–5, 163–4, 165–8
in Uganda 16, 18–19, 170–5
and women 39, 81–2
Lane, C 163–4
law *see* legislation
legislation
  land tenure 11–13, 17–19, 56–7, 73–7, 89, 94, 116
livestock
  alienation 63–4
  cattle raids 41, 151, 170, 184
  commercial breeding 176–7
  development 79–81, 99–102, 157–8, 159
  disease 74, 83, 124, 151, 163, 176
  and the environment 66–7, 107–8, 109
  ownership 5, 36, 82, 83, 102–3, 105–6
Livestock II 79, 81
local authority 17, 18, 43, 95, 110–11, 113–14
Lutheran World Federation 182

Maasai
  and immigration 40–1
  in Kenya 29, 30–1, 36–8, 39–41, 118
  land tenure 36–7, 39–40
  stock associateship 38
  in Tanzania 158, 160, 161, 164
Mali
  aid 61
  colonialism 17, 53–6

conflicts 63–6
conservation 67, 69
development policy 53–63
Dina system 17, 50, 51–3, 55
donors 61
and the environment 66–7
fiscal policy 57, 58–63, 67
flooding 49, 70
food production 60
forestry policy 60, 62, 69
government 50–8
independence 56–8
industry 47
land policy 58–63, 67–70
land tenure 17, 50–8
livestock 46
nationalisation 54, 56–7
population growth 47
rainfall 47, 49
religion 50
resource management 50–63
sedentarisation 51
tribes 49–50
water supply 47, 49, 70
wildlife 49
management
  by pastoralists 11–12, 68–70, 76–7, 100–1, 112, 118, 167–8
  rangeland 13–14, 21–5, 139, 156–60, 170, 180–1
  of resources 8–9, 21–5, 50–63, 68–70, 167–8, 184–5
Manantali dam 73, 87, 110
Mauritania
  aid 79–81
  animal husbandry 71, 74, 88
  case studies 86–8
  colonialism 72–3
  conflicts 82–4, 88–90
  development policy 74–81, 91–2

donors 79–81
and the environment 85–6
food production 74, 77, 88
forestry policy 76
government 72–3
independence 73
industry 73
land policy 18, 73–9
land tenure 18, 20, 72–3, 78–9, 86–8
livestock 5, 88
livestock development 74, 76–7, 79–81
livestock ownership 82, 83
nationalisation 73–4
pastoral organisations 76–7, 90
privatisation 73
projects 79–81
rainfall 80, 85
religion 9, 72, 73–4, 78–9, 88
sedentarisation 72, 80, 83–4
tribes 87
water supply 73, 74, 80
women 81–2
migration 30–1, 74, 124, 131–3 *see also* transhumance
minerals 108
mining 73
mobility 52–3, 100, 143, 144

NARCO 159–60
national parks
  in Kenya 16, 31, 117
  in Senegal 99, 109, 110, 117
  in Tanzania 163
  in Uganda 16, 175–6, 177, 179, 183, 184
nationalisation 9–10, 14–15
  in Mali 54, 56–7
  in Mauritania 73–4
  in Senegal 94
  in Sudan 134–5
  in Tanzania 152–3, 155

in Uganda 173
NGOs 19, 79, 92, 165, 176, 177
Niane, IC 115
Niang, M 113
Niger 12
Niger delta *see* Mali
NOPA 44
NORAD 19, 35
Nouakchott Municipality 79
NPAs 80
Nyerere, Julius 151–2, 154, 159

ODR 61–3
OPEC 79
organisations
  pastoral 24–5, 76–7, 79–80, 83, 90, 91, 115–16
overgrazing 40, 85, 107, 179, 180–1, 184
Oxfam 19, 44, 139, 147, 182

Parkipuny, ML 163, 164
participation 12, 91, 100–1, 140–1, 147–8, 149, 168
PAs *see* organisations, pastoral
Pastoral Tenure Network 44
pastoralism
  and conservation 16, 31, 163, 175–6, 179, 183
  and cultivation 64, 88–9, 109–12, 136–7, 155–6, 160–3, 176–7
  economic contribution 74, 136, 146, 150–1, 164, 184
  management 11–12, 68–70, 76–7, 100–1, 112, 118, 167–8
  'new' 63, 72, 78, 83, 91, 139, 145, 178
  organisations 24–5, 76–7, 79–80, 83, 90, 91, 115–16
  prejudice 82, 103, 136, 155–8, 174–5
  productivity 95–6, 118
  regulated 51–3, 55–6, 60–2, 98–9

representation 20–1, 66, 114–15, 147–8, 165, 183–4
research 43–5, 117–18, 146, 148–9, 182–3, 186–91
survival strategies 38–9, 82–3
PDESO 101, 114
Peuhl 94, 102–3, 108, 110, 112, 114–15
population growth 21, 44, 47
power relations 20–1
prejudice 82, 103, 136, 155–8, 174–5
privatisation 6, 8, 13–14, 15
  in Kenya 36–7
  in Mauritania 73
  in Sudan 135
  in Tanzania 156, 160–1
  in Uganda 172, 173–4
productivity 95–6, 118
projects 79–81, 99, 100–2, 147–8, 159–60, 176
property *see* land
PSAEAS 101, 102

Quam, MD 182

Raintree, J 43
ranch system
  in Kenya 13, 32, 36–7, 37–40
  in Tanzania 157, 159–60
  in Uganda 175, 176–7, 178, 183
rangeland management 13–14, 21–5, 95–6, 139, 156–60, 170, 180–1
reafforestation 59, 109
refugees 179, 180
registration *see* titling
regulating pastoralism 51–3, 55–6, 60–2, 98–9
religion 9, 50, 72, 73, 78–9, 88, 136
representation 20–1, 66, 114–15, 147–8, 165, 183–4
research 43–5, 117–18, **146, 148–9,** 182–3,

186–91
resources
  access to 7, 9–15, 16, 51–2, 103–6
  management 8–9, 21–5, 50–63, 68–70, 167–8, 184–5
rights
  grazing 31, 33, 76, 104–5, 179, 180–1, 183
  tree tenure 19, 34–5
  women's 39, 81–2, 182
Rimaïbé 50, 51, 52, 53–4
rural councils 95, 96, 115, 140–1, 147
Rwanda 14

SAED 98
Sahel 47, 49
Santoir, C 105–6
sedentarisation 10–11
  and the environment 11, 86, 143, 180
  in Mali 51
  in Mauritania 72, 80, 83–4
  in Sudan 137, 146
  in Tanzania 155, 157
Senegal
  aid 99–101
  alienation 102–6
  animal husbandry 102
  case studies 110–13
  colonialism 16–17, 113
  conflicts 20, 102–6, 110–16
  conservation 98–9, 117
  development policy 96, 98–102
  donors 99–101
  and the environment 106–10
  forestry policy 104–5, 111–12
  game reserves 99
  grazing rights 104–5
  land categories 94–5
  land policy 18, 94–6, 98–102, 117
  land tenure 16–17, 18, 94, 110–13, 116
  livestock development 96, 98, 99–102

  livestock ownership 5, 102–3, 105–6
  national parks 99, 109, 110, 117
  nationalisation 94
  projects 99, 100–2
  rainfall 106
  tribes 94
  water supply 100, 103–4, 106
  women 102
settlement *see* sedentarisation
Sharia 72, 73–4, 136
sheep 121–2, 123, 131–2, 133
social issues 35–6, 38–40, 44, 74, 83, 178
SODESP 100, 102, 103–4
'stock associateship' 38
Sudan
  aid 138–9
  alienation 134
  case studies 145–6
  colonialism 16, 134, 140
  Condominium 134
  conflicts 124, 139–41, 147–8, 149
  development policy 137–9
  donors 138–9
  and the environment 123–4, 142–5
  government 134–5, 140–1
  independence 134, 140–1
  land policy 17–18, 21
  land tenure 16, 122–3, 124, 133–7, 141, 146, 148–9
  livestock 5, 120–2, 123
  nationalisation 134–5
  population 121
  privatisation 135
  projects 147–8
  rainfall 123
  religion 136
  sedentarisation 137, 146
  tribes 124, 131–3
  water supply 131–2, 138, 143
  women 139

survival 38–9, 82–3
sustainability 43, 67, 143, 144, 170

Tanzania
  aid 159–60
  case studies 163–4
  colonialism 152–3
  conflicts 160–2, 164–6
  conservation 162, 163
  development policy 158–60
  donors 159–60
  and the environment 20, 162–3
  game reserves 163
  government 151–5, 165–6
  independence 153–5
  land policy 15, 21, 176–7
  land tenure 19, 151–5, 163–4, 165–8
  livestock development 157–8, 159
  livestock ownership 5, 150
  national parks 163
  nationalisation 152–3, 155
  privatisation 156, 160–1
  projects 159–60
  ranch system 157, 159–60
  sedentarisation 155, 157
  tribes 151
  'villagisation' 10, 11, 24, 157
  water supply 161
  women 161–2
Textile Fibre Development Agency 101
Thailand 13–14
titling 11–13, 44, 56–7, 167–8, 174, 179
Touré, O 114
'tragedy of the commons' 5–6, 8, 9–10, 43, 68, 156
transhumance 34, 49, 52, 55, 180 *see also* migration
tree tenure 19, 34–5
tribes
  conflicts 84, 124, 140–1, 147, 164–5, 166–7
  in Kenya 29, 33–7
  in Mali 49–50
  in Mauritania 87
  in Senegal 94
  in Sudan 124, 131–3
  in Tanzania 151
  in Uganda 170–2, 175
Turkana 19, 29, 34–6, 41
Turkwel dam 34, 35

Uganda
  aid 176
  animal husbandry 176
  case studies 182–3
  cattle corridor 170, 176, 181
  climate 175, 180
  colonialism 16, 172–3
  conflicts 178–80, 183–4
  conservation 16, 175–6, 177, 179, 184
  development policy 175–7
  donors 176
  and the environment 169–70, 177, 180–1, 184
  fiscal policy 172
  game reserves 16, 176, 177, 179, 183, 184
  government 170–5, 183
  independence 173–5
  industry 177
  land policy 18–19, 183–4
  land tenure 16, 18–19, 170–5
  livestock 170
  livestock development 176–7
  national parks 175–6, 177, 179, 183, 184
  nationalisation 173
  privatisation 172, 173–4
  projects 176
  rainfall 175
  ranch system 175, 176–7, 178, 183
  refugee settlement 179, 180

sedentarisation 10
social issues 178
tribes 170–2, 175
water supply 177
women 181–2, 184
UN Development Programme 81
UN Food Security Commission 81
UNCED 25
UNICEF 44
United Nations 139, 147 *see also* UN
UNRISD 2
USAID 19, 159–60, 176, 177
use of land 1–2, 8, 10–13
usufruct 1, 34, 74, 83, 105, 135, 136

'villagisation' 10, 72, 80, 157
VPAs 80

water
  accessibility 9, 13, 33–4, 88–9, 100, 103–4
  conflicts 41, 88–9, 110
  and the environment 20, 66–7, 85, 143, 177, 181
wells 80, 81, 86, 89, 161
West African Economic Community 81
women
  activity 82, 161–2, 181–2
  changing lifestyle 3–4, 25, 39, 139, 161–2
  cooperatives 81–2
  and land tenure 39, 81–2
  and livestock 82, 102, 181–2, 184
  and resources 4
  rights 39, 81–2, 182
  vulnerability 4
World Bank 11, 76, 79, 81, 159–60

OTHER RELEVANT PUBLICATIONS AVAILABLE FROM EARTHSCAN

Social Change and Conservation
Edited by Krishna B Ghimire and Michel Pimbert

Protected areas and conservation policies are usually established with only local nature and wildlife in mind. Yet they can have far-reaching consequences for local populations – often harmful ones, undermining their access to resources and their livelihoods. This is the first fully comprehensive discussion of the social consequences of protected area schemes and conservation policies. Drawing on case studies from North America, Europe, Asia, Central America and Africa, it critically reviews current trends in protected area management and the prevailing concept of conservation, and shows how local people have been affected – their customary rights, livelihoods, wellbeing and social cohesion. Lack of local participation, of respect for local rights, and too much emphasis on market forces has usually meant the failure to provide for human concerns and wellbeing. The leading authorities in this book – including Marcus Colchester, Piers Blaikie, Michel Pimbert and Jules Pretty – argue for a thorough overhaul of current conservation thinking and practice.

Published in association with UNRISD
£18.95 paperback   ISBN 1 85383 410 6   224pps   available now

Conservation Through Survival
Indigenous Peoples and Protected Areas
Stan Stevens

For more than a century the establishment of national parks and protected areas was a major threat to the survival of indigenous people. The creation of parks based on wilderness ideals outlawed traditional ways of life and forced from their homelands communities which had shaped and preserved local ecosystems for centuries.

Today such tragic conflicts are being superseded by new alliances for conservation. *Conservation Through Survival* assesses cutting-edge efforts to establish new kinds of parks and partnerships with indigenous peoples. It chronicles new conservation thinking and the establishment around the world of indigenously-inhabited protected areas; provides detailed case studies of the most important types of co-managed and indigenously managed areas; and offers guidelines, models and recommendations for international action.

Contributors who have been actively involved in projects around the world provide in-depth accounts from Nepal, Australia, New Guinea, Nicaragua, Honduras, Canada and Alaska. This book will be required reading for environmentalists, protected area planners and managers, and all who care about the future of indigenous peoples and their homelands.

£17.95 paperback   Island Press ISBN 1 55963 449 9   320pp   available now

Farms, Trees and Farmers
Responses to Agricultural Intensification
Edited by JE Michael Arnold and PA Dewees

'a compendium of knowledge and experience... that is a unique contribution to the literature on farm forestry... It should be required reading for all students and practitioners concerned with the subject' Peter Wood, Scottish Forestry

This book is a milestone in understanding the role of trees grown, outside forests, on farms throughout the developing world, particularly in the light of increasing intensification of agriculture. It examines subsistence needs, market opportunities and constraints, allocation of land and labour, and risk management. In showing how farmers decide to grow or not to grow trees, it fills a gap in our knowledge about farming systems and provides a guide to encouraging farm forestry throughout the world. This text will be of interest to students and professionals in forestry, agriculture, rural development and conservation.

£16.95 paperback   ISBN 1 85383 484 X   304pp   available now

Sustaining the Soil
Indigenous Soil and Water Conservation in Africa
Edited by Chris Reij, Ian Scoones and Camilla Toulmin

'I strongly recommend this book, not only because it provides interesting information about a wide range of situations and locally developed or adapted technologies, but also because it provokes reflection, and provides a basis for better mutual understanding and partnerships between farm families and technicians' TF Shaxson

'It should be compulsory reading for all persons engaged in sustainable agriculture, especially those who prefer the quick technical fix, and who regard small farmers as ignorant peasants' Michael Stocking, Professor of Natural Resource Development, University of East Anglia

For the first time on an Africa-wide scale, this book shows that indigenous techniques work, and suggests new ways forward for governments and agencies attempting to support sustainable land management in Africa.

Published in association with IIED
£12.50 paperback   ISBN 1 85383 372 X   240pp   available now

Regenerating Agriculture
Policies and Practice for Sustainability and Self-Reliance
Jules N Pretty

'This is the book we have so badly needed... It should be read and acted on by all concerned with agricultural policy, research and extension' Dr Robert Chambers

'an opportunity to see where practice and experience are often ahead of theory' Professor Norman Uphoff, Director, CIFAD, Cornell University

'Wonderful, extremely well written. It blends analysis, argument and example very well. This book will challenge you' Ed Mayo, New Economics Foundation

'Sure to become an instant classic. It will bridge the gap between technical and social perspectives' Pesticides News

Regenerating Agriculture looks at the scale of the challenge facing agriculture today and details the concepts and characteristics of alternative, sustainable agricultural practices. Jules Pretty draws together new empirical evidence from a diverse range of agro-ecological and community settings to show the impacts of more sustainable practices. Using 20 detailed case studies, and field and community-level data from more than 50 projects and programmes in 28 countries, he identifies the common elements of success in implementing sustainable practices and shows how to replicate them. In addition, he looks at the existing policy frameworks and institutional processes, and sets out 25 alternative policies which are known to work to support the shift to greater self-reliance and sustainability in agriculture.

£12.95 paperback ISBN 1 85383 198 0   £29.95 hardback ISBN 1 85383 227 8   320pp available now

For details of these and other Earthscan publications, and for a copy of our catalogue, please contact:

Earthscan Publications Ltd
120 Pentonville Road
London N1 9JN
Tel: 0171 278 0433
Fax: 0171 278 1142
email: earthinfo@earthscan.co.uk
website: http:\\www\earthscan.co.uk